SpringerBriefs in Ethics

More information about this series at http://www.springer.com/series/10184

Ellen-Marie Forsberg · Clare Shelley-Egan
Erik Thorstensen · Laurens Landeweerd
Bjorn Hofmann

Evaluating Ethical Frameworks for the Assessment of Human Cognitive Enhancement Applications

 Springer

Ellen-Marie Forsberg
Work Research Institute, Research Group
  on Responsible Innovation
Oslo and Akershus University College
Oslo
Norway

Laurens Landeweerd
Work Research Institute, Research Group
  on Responsible Innovation
Oslo and Akershus University College
Oslo
Norway

Clare Shelley-Egan
Work Research Institute, Research Group
  on Responsible Innovation
Oslo and Akershus University College
Oslo
Norway

Bjorn Hofmann
Norwegian University of Science
  and Technology
Gjøvik
Norway

Erik Thorstensen
Work Research Institute, Research Group
  on Responsible Innovation
Oslo and Akershus University College
Oslo
Norway

ISSN 2211-8101   ISSN 2211-811X   (electronic)
SpringerBriefs in Ethics
ISBN 978-3-319-53822-8   ISBN 978-3-319-53823-5   (eBook)
DOI 10.1007/978-3-319-53823-5

Library of Congress Control Number: 2017932632

Printed on acid-free paper

This Springer imprint is published by Springer Nature
The registered company is Springer International Publishing AG
The registered company address is: Gewerbestrasse 11, 6330 Cham, Switzerland

Acknowledgements

The research leading to these results has received funding from the Norwegian Financial Mechanism 2009–2014 and the Ministry of Education, Youth and Sports of the Czech Republic under Project Contract no. MSMT-28477/2014, Project no. 7F14236. We would like to thank the anonymous reviewer for helpful suggestions for improving the book and the series editors for good collaboration.

Acknowledgments

Contents

Abstract

Human cognitive enhancement (HCE) is a term signifying applications that are supposed to improve cognitive capacities, such as attention, memory or reasoning. Such enhancement can be carried out in various ways through the use of pharmaceuticals, genetic interventions, brain stimulation, sensors, and other tools, such as smart glasses. As the ethical debate about HCE has been rather polarised, the aim of this book is to make a reasoned argument for sound ethical frameworks that might be used by decision-makers to ethically assess HCE applications in a pluralist and nuanced way. We do this by considering a selection of ethical frameworks used in fields closely adjacent to HCE and evaluate their applicability to ethical decision-making within the HCE field. We argue that appropriate frameworks should be able to incorporate all relevant ethical values (i.e. be comprehensive), facilitate transparency and be user-friendly for non-philosophers. The ethical discussions concerning two HCE applications—cognitive enhancing drugs and non-invasive brain stimulation—are analysed in order to generate an overview of ethical issues that the frameworks must be able to encompass. Our evaluation shows that principle-based approaches appear to be the best candidates for such an ethical framework, and we develop variants of the ethical matrix and the ethical impact assessment approach for use in the HCE field.

Keywords Human enhancement · Human cognitive enhancement · Ethical framework · Pharmaceutical cognitive enhancers · Non-invasive brain stimulation

Chapter 1
Introduction

1.1 About Human Cognitive Enhancement, the Philosophical Debate and the Need for a Practical Ethical Framework for Assessing HCE Applications

Human enhancement (HE) is the common denominator for applications or activities that are designed to temporarily or permanently improve human beings in different ways, as opposed to merely repairing damages. While therapy is often defined as "the attempt to restore a certain condition (e.g. normality, sanity, health)", enhancement is viewed as transcending these boundaries (STOA 2009, 10). Human cognitive enhancement (HCE) is a term that signifies applications that are supposed to improve cognitive capacities, such as attention, memory or reasoning. Such enhancement can be carried out in various ways; through the use of pharmaceuticals, genetic interventions, brain stimulation, sensors, and other tools, such as smart glasses. There is no agreed-upon technological range of what may be labelled human cognitive enhancement applications; the category is rather defined by the purpose for which interventions or applications are applied, namely enhancement.

The notion of enhancement is controversial and the debate about enhancement has been quite polarised. "Transhumanists", such as Nick Bostrom or Julian Savulescu, champion HE in principle, adhering to the idea that our given nature is in some sense a restriction. Nature has brought us quite a distance in our evolution, but there are also many flaws in our nature, as well as many potentials that are not fully developed. In their view, human enhancement would be an instrument to rid us of our natural chains, enable us to take a step away from natural evolution, and enter a more controlled and accelerated post-evolutionary, or techno-evolutionary process. In some cases, this step is not framed as merely a positive trigger, but even as a moral duty (Harris 2007).

The terminology is not neutral here; labelling an effect as an enhancement provides a positive framing. However, for some, human enhancement also carry along negative aspects such as the idea that enhancement is not natural, or the

© The Author(s) 2017
E.-M. Forsberg et al., *Evaluating Ethical Frameworks for the Assessment of Human Cognitive Enhancement Applications*, SpringerBriefs in Ethics, DOI 10.1007/978-3-319-53823-5_1

concern that in enhancing our capabilities, we instrumentalise ourselves. Moreover, the term enhancement may imply that we know and agree on what would count as being 'enhanced', but authors such as McGee (1997) have a more differentiated view. It is argued that enhancement cannot simply be regarded as intrinsically positive. In many cases, the amelioration of one capability or characteristic may carry along a diminishment of other capabilities. Thus, an enhancement of mood might carry along a diminishment of certain cognitive abilities, and the enhancement of, for example, creativity may diminish one's ability to concentrate (see Landeweerd 2010).

Other concepts also frame the debate. For instance, the "transhumanists" have labelled their opponents as "bio-conservatives". This diverse group of scholars[1] argue that human dignity is at stake and that human enhancement is unnatural or inauthentic for the human race. For instance, Kass (2003) says:

> Most of the given bestowals of nature have their given species-specified natures: they are each and all of a given *sort*. Cockroaches and humans are equally bestowed but differently natured. To turn a man into a cockroach—as we don't need Kafka to show us—would be dehumanizing. To try to turn a man into more than a man might be so as well. We need more than generalized appreciation for nature's gifts. We need a particular regard and respect for the special gift that is our own given nature (2003, 1).

Being similarly concerned by our possibility to change human nature, Fukuyama says:

> Denial of the concept of human dignity – that is, of the idea that there is something unique about the human race that entitles every member of the species to a higher moral status than the rest of the natural world – leads us down a very perilous path. We may be compelled ultimately to take this path, but we should do so only with our eyes open. (2002, 149)

Different concepts are interrelated with enhancement in general and cognitive enhancement specifically. These include, from a more ideological point of view, the debates on posthumanism (where Peter Sloterdijk has been a prominent figure) and the before mentioned transhumanism (where prominent proponents include Nick Bostrom, Julian Savulescu and Anders Sandberg). Moreover, an important concern regarding human enhancement is its implicit association with, and negative connotation of, the 'superhuman' (Übermensch), which relates to positivistic claims regarding genetic amelioration and the eugenics movements of the early and mid 20th century (Habermas 2003). Enhancement is also associated with the computer metaphor of a human 'upgrade'.[2] Through all these definitions and associated debates, concepts of human nature play an important implicit normative role.

This strongly framed debate (particularly prominent at the start of the millennium) has been quite polarised. As Selgelid observes, "parties to the debate risk talking past one another" (2008, 237). The ethical debate in the area has also been

---

[1]Bostrom (2005) includes the following in this category: Leon Kass, Francis Fukuyama, George Annas, Wesley Smith, Jeremy Rifkin, and Bill McKibben.

[2]See for instance http://www.theatlantic.com/technology/archive/2012/02/more-than-human-the-ethics-of-biologically-enhancing-soldiers/253217/. Accessed 1 May 2016.

haunted by the lack of empirical basis, while "(…) the meaning, consequences and ethics of enhancement" largely turn on key empirical questions (ibid, p. 238). As Outram (2012) observes, several commentators have argued that ethics is being led into "unwarranted territory" with little scientific and sociological empirical evidence to support underlying claims regarding efficiency and usage of existing technologies (p. 173).

There have therefore recently been calls for ethically-sensitive policy-making and appropriate governance approaches in the area of human enhancement (STOA 2009; Zwart 2015). Selgelid (2008) observes that a well developed ethical framework that can deal with conflicting values such as liberty, equality and utility, and the need to strike a balance between them, is lacking. Schermer et al. (2009) point out that much of the debate has centred on "enhancements with a capital E", i.e. those enhancements that go beyond what we currently understand to be "normal" or naturally human, and argue that there are a number of important ethical and policy questions to be addressed concerning less spectacular kinds of enhancement—or "enhancements with a small e"—that are already feasible or will be in the near future.

For public policy- or decision-making, this is important. The transhumanist/ bioconservative debate does not encourage the identification of common ground and in pluralist societies simply picking one favoured philosophical stance is not defensible as a basis for societal decision making (Forsberg 2007). For such policy or decision making, more practical ethical approaches would seem to be required, where there is at least a potential link to evidence to be gathered in the short or medium term and which includes a broader range of value bases that would at least in principle allow convergence on certain judgements. As a basis for policy or decision making Toulmin (1981) argues that staying at a principled level is likely to cement differences, while operating at a common morality based middle level is more likely to facilitate a moral dialogue that may agree on important points (in spite of ideological differences). This might be a viable strategy for furthering the discussion on HE beyond the dead-lock of transhumanists and bioconservatives.

Indeed, pharmaceutical and technological advances make the discussion on HCE less speculative and more urgent. For example, transcranial direct current stimulation (tDCS) for cognitive enhancement in healthy individuals is becoming increasingly popular. tDCS devices can be constructed at home using straightforward instructions and cheap parts (Fitz and Reiner 2013). At the same time, the reach of transcranial direct current stimulation is extending beyond home users, with companies selling compact and user-friendly devices (Farah 2015). Similarly, pharmaceutical cognitive enhancers which work to enhance certain cognitive functions are said to be widely used by students and academics (Greely et al. 2008). The Nuffield Council (2013) notes that the number of potential users for the non-therapeutic application of novel neurotechnologies—including tDCS—is inevitably much greater than that for specialised medical applications, thus any ethical or social concerns that do arise warrant attention (p. 163). Moreover, the size and nature of the market for non-therapeutic applications "(…) raises the prospect of direct to consumer (DTC) marketing of devices and services and private use of neurotechnologies unmediated by healthcare professionals" (p. 163). The same is true for pharmacological enhancers, where it is claimed that off-label prescription[3] of medication for explicit enhancement purposes is likely to expand (Schermer et al. 2009).

In addition to bringing to the fore questions about fundamental cultural values and what it means to be human, these HCE applications raise questions regarding the cost of the applications, issues of distributive justice and the applications' societal desirability. Moreover, the complexity of brain function and its effect on behaviour and cognition raises questions of the effectiveness of interventions and of unintended side effects (Chan and Harris 2006). For practical decision making on HCE such issues are just as important as the more philosophical ones.

The purpose of this book is thus to offer a proposal for how to address such a variety of ethical concerns and arguments in the HCE field, going beyond the polarised philosophical debate, and contributing to more concrete and nuanced ethically sensitive decision making by private and public decision makers on specific HCE applications. We will begin by introducing the notion of ethical tools and review the discussion of such tools in neighbouring fields of emerging technologies. This will provide us with a basis for evaluating different possible approaches to ethical assessment of HCE applications, and, ultimately, allow us to propose recommendations for such assessment.

References

Bostrom, N. 2005. In defense of posthuman dignity. *Bioethics* 19 (3): 202–214.
Chan, S., and J. Harris. 2006. Cognitive regeneration or enhancement: The ethical issues. *Regenerative Medicine* 1 (3): 361–366.
Farah, M.J. 2015. The unknowns of cognitive enhancement. Can science and policy catch up with practice? *Science* 350 (6259): 379–380.
Fitz, N.S., and P.B. Reiner. 2013. The challenge of crafting policy for do-it-yourself brain stimulation. *Journal of Medical Ethics* 0: 1–3. doi:10.1136/medethics-2013-101458.
Forsberg, E.-M. 2007. Pluralism, the ethical matrix and coming to conclusions. *Journal of Agricultural and Environmental Ethics* 20 (5): 455–468.
Fukuyama, F. 2002. *Our Posthuman Future*. London: Profile.
Greely, H., B. Sahakian, J. Harris, R.C. Kessler, M. Gazzaniga, P. Campbell, and M.J. Farah. 2008. Towards responsible use of cognitive-enhancing drugs by the healthy. *Nature* 456: 702–705.
Habermas, J. 2003. *The Future of Human Nature*. Polity Press.
Harris, J. 2007. *Enhancing Evolution. The Ethical Case for Making Better People*. Princeton University Press.
Kass, L.R. 2003. Ageless bodies, happy souls. *The New Atlantis* 1: 9–28.
Landeweerd, L. 2010. *Asberger's Syndrom, Bipolar Disorder and the Relation between Mood, Cognition, and Well-Being*, ed. J. Savulescu, R. ter Meulen, and G. Kahane, 207–217. Wiley-Blackwell.

---

[3]Off-label use refers to "the prescription of drugs for a purpose that is not included in the drug's approved label" (Schermer et al. 2009, p. 78).

McGee, G. 1997. Parenting in an era of genetics. *Hastings Center Report* 27 (2): 16–22.

Nuffield Council on Bioethics. 2013. *Novel neurotechnologies: Intervening in the brain.* London: Nuffield Council on Bioethics.

Outram, S.M. 2012. Ethical considerations in the framing of the cognitive enhancement debate. *Neuroethics* 5 (2): 173–184.

Schermer, M., I. Bolt, R. de Jongh, and B. Olivier. 2009. The future of psychopharmalogical enhancements: Expectations and policies. *Neuroethics* 2 (2): 75–87.

Science and Technology Options Assessment (STOA). 2009. In *Human enhancement study (IP/A/STOA/FWC/2005-28/SC35, 41 and 45)*, ed. C. Coenen, M. Schuijff, M. Smits, P. Klaassen, L. Hennen, M. Rader, and G. Wolbring. https://www.itas.kit.edu/downloads/etag_coua09a.pdf. Accessed October 22, 2015.

Selgelid, M.J. 2008. Advancing posthuman enhancement dialogue. In *Medical Enhancement and Posthumanity*, eds. B. Gordijn and R. Chadwick, 237–40, Springer.

Toulmin, S. 1981. The tyranny of principles. *The Hastings Center Report* 11 (6): 31–39.

Zwart, H. 2015. *Deliverable D3.5 Final Report. Neuro-enhancement Responsible Research and Innovation* (NERRI). A project funded by the European Commission.

Chapter 2
Ethical Governance and Ethical Tools

Ethically sensitive decision making is needed both with regard to general policies on human enhancement and with regard to specific enhancement applications. Science, technology and innovation policies may, on the one hand, be developed to support or steer HE technology trajectories in certain directions. This requires societal deliberation regarding the kind of innovation we, as a society, want to encourage. For this purpose, it could be important to consider, for instance, whether the technology inherently contributes to or challenges our concepts of agency, autonomy or personhood. This is the kind of deliberation that Stirling (2008) presents as "opening up" reflection, questioning both the purposes of the innovations and considering alternative ways of achieving these purposes (including non-technological means) (Rip and te Kulve 2008). An informative approach to deliberating the overall ethical and philosophical questions (the Socratic Health Technology Assessment) has been developed by Hofmann (2016), Hofmann et al. (2016) based on a review of Health Technology Assessment approaches.[1]

As we argue in Chap. 1, there is, on the other hand, also a need for ethical assessment of specific applications. This kind of ethical assessment will often have a regulatory or decision-making focus. Risk assessments are commonly applied in the area of pharmaceuticals or medical inventions, but sometimes there may also be a desire to ethically assess specific applications.[2] For instance, potential consumers or users (such as socially responsible doctors) may want to ethically assess HE applications before they make purchasing-decisions; producers may want to ensure that their products are ethically acceptable; or policy makers may want an ethical

---

[1]Hofmann's work was somewhat similar to the work presented in this book, but took as a starting point the Health Technology Assessment tradition, while we in this book specifically consider the tradition of practical ethical frameworks in emerging technologies. The two strands of work also target different decision making levels.

[2]In Norwegian gene technology legislation such application focused ethical assessments are in fact required for all genetically modified organisms (GMOs) as a supplement to scientific risk assessment.

© The Author(s) 2017
E.-M. Forsberg et al., *Evaluating Ethical Frameworks for the Assessment of Human Cognitive Enhancement Applications*, SpringerBriefs in Ethics, DOI 10.1007/978-3-319-53823-5_2

assessment when considering policy interventions. This is the kind of deliberation that Stirling (2008) describes as "closing down". This is the decision making level targeted in this book, where no framework yet has been proposed in the field of HE.[3]

These opening up and closing down levels will be interrelated, even if the two different kinds of deliberation have different characteristics. Societal deliberation will inform the frameworks that guide application oriented decision making, and vice versa, and both may inform policy, albeit in different ways. The philosophical debate between the transhumanists and the bioconservatives may be most valuable for informing more open deliberation of HE technologies because they raise general philosophical questions. However, the strongly framed philosophical stances are not apt as a basis for regulatory or decision oriented policy-making, as these positions are hard to reconcile with the current paradigm of evidence-based policy making.[4]

2.1 Ethical Tools as Vehicles for Practical, Ethical Dialogue

Beekman and Brom (2007) suggest in the biotechnology context the use of "practical instruments that can be used (tools) in order to support debates and deliberative structures for a systematic engagement with ethical issues" (2007, 4).[5] Beekman and Brom refer to a European project, Ethical Biotechnology Assessment Tools (the Ethical Bio-TA Tools project), which identified and reviewed a series of ethical tools or frameworks. The Ethical Bio-TA Tools project identified three major types of tools: Decision-making tools, public consultation and involvement tools and food chain value communication tools (see Table 2.1).

Decision-making tools were defined as tools that would aid ethical decision- and policy-making, in other words, assist in closing down on decisions. Such tools should not be seen as mechanical decision-processes, but rather as "something that can help you use your judgement" (Seedhouse 2009, 107). *Public consultation* tools designated *process* tools with procedures to elicit information on facts and values from experts, stakeholders or lay people, as well as procedures to deliberate on these. Examples are citizen's panels, Delphi processes, stakeholder workshops and consensus conferences (see Fixdal 2003; Rowe and Frewer 2000). These public consultation tools mainly have the purpose of opening up debates. *Food chain*

---

[3]Indeed, from Forsberg et al. (2014, see Table 5.1) it appears that this is a general gap in the ethical assessment of emerging technologies.

[4]It should be noted that proponents of these approaches present a rich variety of ethical arguments, not only the abstract, philosophical ones that seem to lead to deadlocks. These more concrete arguments will be important to consider in any ethical discussion of HCE.

[5]In this book, the terms "tools" and "frameworks" will be used interchangeably.

Table 2.1 List of frameworks discussed in the ethical bio TA tools project	*Decision-making framework*
	Casuistry
	COGEM framework
	Critical systems heuristics
	Delphi method
	Discourse ethics
	Ethical codes/guidelines
	Ethical matrix
	Multi-criteria mapping
	Precautionary principle
	Principle based ethics
	Risk analysis
	Stakeholder analysis
	Value-tree analysis
	Public consultation and involvement
	Citizen's forum
	Consensus conference
	Focus group
	Future workshop
	Public hearing
	Public forum
	Referendum
	Scenario workshop
	Technology Delphi studies/technology foresight
	Food chain value communication
	Benchmarking
	Ethical accounting
	Ethical audits
	Ethical codes
	ISO 9000
	Normative standards
	Stakeholder dialogue
	Stepwise dilemma-solving
	Total quality management
	Value clarification
	Weston's toolbox

value communication tools were especially targeted to the needs of stakeholders in an industrial context.

We believe in general that the ethical discussion in HCE can benefit from systematic reflection on ethical tools, in a manner similar to the biotechnology context. We can therefore learn from the Ethical Bio TA Tools project. More specifically, we argue that the decision making tools are generally appropriate for application oriented decision/policy making needs, which is our focus in this book. Public consultation tools are especially important for societal deliberation of technology policy in general. Finally, food chain value communication tools may be relevant as tools for the (future) HCE industrial value chains, but will not be addressed here.

2.2 Quality in Practical Ethics

For all of these kinds of ethical assessment tools, a crucial concern is to ensure their quality. An ethical assessment tool is of no value if it is of low quality. Ethical assessments should not be biased or misleading, or refer to non-relevant issues. Such concerns with quality have led researchers to study what characterises good ethical assessments. In particular there has been a focus on how ethical tools may help to ensure high quality ethical assessment. In the Ethical Bio-TA Tools project, the term "soundness" was used to indicate a concern for methodological quality of ethical frameworks:

> an ethical framework is ethically sound *iff* [if and only if] its application produces understanding of ethically relevant considerations in such a way that within a given body of knowledge and on condition of its competent use no further considerations would decisively alter the normative conclusions drawn from the framework by the users (Kaiser et al. 2004, 26).

Ethical soundness was operationalised as:

1. Inclusion of values at stake;
2. Transparency;
3. Multiplicity of viewpoints;
4. Exposition of case relevant ethically relevant aspects;
5. Inclusion of ethically relevant arguments (Kaiser et al. 2004, 27).

Moula and Sandin (2015) have some objections to Kaiser et al.'s approach, but similarly conclude that all high quality ethical frameworks should be *comprehensive*, i.e. consider a broad range of ethically relevant aspects. They also stress the need for frameworks to be *user-friendly*. For decision guiding tools they also argue that the criteria of transparency, guiding users toward a decision and justification of the decision-supporting mechanism as important.

In bioethics there has similarly recently been an extensive reflection on what makes an ethical analysis or approach good. Even if this reflection has been more related to what good scholarly analysis is rather than what characterises a good, practical ethical tool, it is useful to consider this discourse.

One group of quality criteria addresses the impact of ethical approaches. Rothman (1991) and Caplan (2015) propose that a good approach is one that alters moral authority. Stevens (2003) and Koch (2012) suggests that an ethical approach is good if it sells bioethics to the establishment and generates funding. Fox (2008) and Evans (2011) argue that a good approach provides apparent objective expertise in identifying and addressing ethical issues. Sheehan and Dunn (2013) focus on the approach's ability to influence policy, while Chan (2015) suggests that it should forge workable (policy) decisions by responding to moral disagreement and generating new understandings. Harris (2015) suggests more generally that an ethical approach should make the world a better place.

Sugarman and Sulmasy (2010) suggest achievement-based criteria for goodness. A good approach or framework expands knowledge, exposes hidden assumptions, challenges prevailing convictions, makes rigorous arguments, enriches understanding, and illuminates contentious issues in new ways (Sugarman and Sulmasy 2010). This seems to be a stance very much consistent with Kaiser et al. (2004) and also with Dunne (2012) who focuses on a framework's ability to develop justifiable and practically useful arguments.

Good frameworks may also be characterized by various process characteristics. Bowman claims that medical ethics should actively seek "perspectives and contributions from people other than academics and clinicians" and empower to action (Bowman 2015, p. 61). Montgomery claims that bioethics should focus "more on who does things, how and why they do them, than in what they study and what they conclude." (Montgomery 2016, p. 20). That is, the quality of bioethics depends on the legitimacy of its institutions to operate in the public sphere (depending on how their members are selected; the nature of the authority that they exercise; the processes by which positions are reached; the efficiency, proportionality and effectiveness of the accountability mechanisms that can be invoked), on the forms of institutionalisation of bioethics, as well as on the forms of bioethics governance; such as opinions, reports, guidelines or consensus statements.

Eckenwiler and Cohn (2009) and Hedgecoe (2004) have proposed criteria related to the tasks of the frameworks, such as being social critics or watchdogs, and challenging injustice.

Rhodes is one of the scholars who provides the most specific and operational criteria. She argues that a good framework in ethics "is coherent, illuminating, accurate, reasonable, consistent, informed, and measured" (Rhodes 2015). By this she means that the analysis should be in coherence with existing systems of norms and values in practice, clarify conceptual issues and solve real problems, refrain from simplistic understandings and unreasonable adherence to specific rules or principles, be informed by a broad range of sources, and avoid exaggerations.

Thus, most of the criteria found in bioethics are general criteria related to its process, its tasks (critical or supportive), its overall goal, or its achievements or impacts.

2.3   Criteria for Good Ethical Tools

As outlined above, in this book, we are interested in tools that help us go beyond the philosophical debates between transhumanists and bioconservatives in order to assist practical decision makers in concrete ethical deliberation and decision making. In this sense we would look for achievement-based tools that aim to increase *comprehensiveness* and broad inclusion of values. This is in line with Kaiser et al. (2004), Moula and Sandin (2015), Sugarman and Sulmasy (2010), Dunne (2012) and Rhodes (2015). Moreover, we assume that these practical decision makers are seldom learned philosophers and we therefore attach great importance to the criterion of *user-friendliness*, which is also highlighted by Moula and Sandin (2015). In terms of impact, we would agree with Sheehan and Dunn (2013) and Chan (2015) that a framework should forge workable decisions; indeed, finding out how to do this in a sound way is the point of departure in this book. However, in a pluralist society we do not think that a framework that is designed to produce a certain impact, such as 'selling' HCE or warning against HCE, is an option for decision makers with a public mandate. Rather, for such decision makers the capacity of the framework to make value judgements *transparent* would be more important. This holds especially for public decision makers. In our view, this transparency should include specifically show how users are guided toward a decision and on what basis they reach it.[6] One might perhaps argue that the importance of user-friendliness is already implied in the notion of a 'tool'. A tool must somehow be seen as useful by a user in a certain setting. Since we target a range of users, the tool must have a certain broader user-friendliness.

Moreover, it is hard to conceive that a broadly usable framework generally should aim at changing moral authority (Rothman 1991; Caplan 2015); it will depend on the situation who has moral authority and whether this authority should be changed.

It is not so important in the decision context targeted here to assess the way the governance structure is organised (for instance, the involvement of the lay people in ethics committees). We are here interested in decision making tools that can be used by a variety of actors, and not in a more specific governance setting. We are not so interested in process criteria (Bowman 2015; Montgomery 2016) or process tools (such as consensus conferences) because we target a variety of decision makers and it would be too much to expect, for instance of a small company wanting to develop sensor based HCE, to organise such a process. Such tools are generally used by professional assessment institutions, such as technology assessment (TA) boards. Here we are rather interested in tools that help the user to identify and analyse ethical issues through the provision of information regarding relevant values or principles. Tools with an explicit ethical content may be used by anyone and should help public or private decision makers to communicate their views and judgements. In this situation, epistemological criteria, such as those proposed by Rhodes (2015), Kaiser et al. (2004) and Moula and Sandin (2015) would be more useful. In sum, from reviewing the proposed criteria we seem to end up with comprehensiveness, transparency and user-friendliness as the most relevant criteria.

---

[6]In other words, we combine Moula and Sandin's three specific criteria for decision-guiding tools into the transparency criterio.

It should be clear that there is a need for sound ethical frameworks for both levels of analysis indicated above; for both open ethical deliberation on the technologies policies and developments and ethical assessment that may aid more practical decision- and policy making; and the quality criteria for frameworks at the two levels may be quite similar.[7] However, in this book we focus on ethical frameworks for assessment of specific applications (or generic groups of applications) with a clear decision making focus, for instance related to decisions as to whether or not to buy, market or to allow marketing of such applications. In such a situation, we will take as a starting point the need for frameworks to facilitate transparent ethical decision making in practice and to be usable for non-philosophers. Moreover, according to the soundness definition given above, this framework needs to be able to incorporate the values at stake, a multiplicity of viewpoints, exposition of case relevant ethically relevant aspects and inclusion of ethically relevant arguments (i.e. what we can call comprehensiveness). Regarding the latter two aspects, the framework should provide directions for further fact finding that will clarify the ethical aspects.

The aim of this book is thus to make a reasoned argument for a sound ethical approach that might be used by decision makers to ethically assess HCE in a comprehensive, transparent and user-friendly way. We will do this by systematically considering existing frameworks from adjacent technology fields and determining their fit for HCE related issues. In order to do this we need first to outline HCE related ethical issues, as these concerns must be addressable in the frameworks. This will first be done by presenting two generic cases of HCE applications and their related ethical issues. After presenting these issues we will identify different candidate frameworks for ethical assessment in HCE and evaluate the ways in which these frameworks are able to transparently address the ethical issues while being user-friendly. We will then be in a position to make recommendations about ethical assessment frameworks that will aid users to draw conclusions in a sound way.

In the following section, we will offer a description of the two main application areas of focus here, i.e. pharmaceutical enhancers and non-invasive brain stimulation techniques. These two cases have been chosen because they are already available on the market and will be increasingly available, as we have already described in the introduction. They have also been discussed in the literature. We will list and briefly describe the general ethical issues that are discussed in relation to the two cases. We focus specifically on non-therapeutic applications and refer only to therapeutic applications where ethical issues touch on both kinds of application or where they are mentioned in accounts about non-therapeutic applications. It should be noted that the ethical issues identified for these two cases cover most of the issues that are relevant for HCE applications so the validity of the argument in this book extend beyond these two cases.

---

[7]As indicated in Hofmann (2016).

References

Beekman, V., and F.W.A. Brom. 2007. Ethical tools to support systematic public deliberations about the ethical aspects of agricultural biotechnologies. *Journal of Agricultural and Environmental Ethics* 20 (1): 3–12.

Bowman, D. 2015. What is it to do good medical ethics? Minding the gap(s). *Journal of Medical Ethics* 41 (1): 60–63.

Caplan, A.L. 2015. Done good. *Journal of Medical Ethics* 41 (1): 25–27.

Chan, S. 2015. A bioethics for all seasons. *Journal of Medical Ethics* 41 (1): 17–21.

Dunn, Michael, et al. 2012. Toward methodological innovation in empirical ethics research. *Cambridge Quarterly of Healthcare Ethics* 21 (04): 466–480.

Eckenwiler, L.A., and F.G. Cohn, eds. 2009. *The ethics of bioethics: Mapping the moral landscape.* JHU Press.

Evans, J.H. 2011. *The history and future of bioethics.* New York: Oxford.

Fixdal, J. 2003. *Public Participation in Technology Assessment.* Ph.D. thesis, Universitetet i Oslo.

Forsberg, E.-M., E. Thorstensen, R.Ø. Nielsen, and E. de Bakker. 2014. Assessments of emerging science and technologies: Mapping the landscape. *Science and Public Policy* 41 (3): 306–316.

Fox, R., and J. Swazey 2008. *Observing bioethics.* New York: Oxford.

Harris, J. 2015. What is it to do good medical ethics? *Journal of Medical Ethics* 41 (1): 37–39.

Hedgecoe, Adam M. 2004. Critical bioethics: Beyond the social science critique of applied ethics. *Bioethics* 18 (2): 120–143.

Hofmann, B. 2016. Toward a method for exposing and elucidating ethical issues with human cognitive enhancement technologies. *Science and Engineering Ethics.* doi:10.1007/s11948-016-9791-0.

Hofmann, B., D. Haustein, and L. Landeweerd. 2016. Smart-glasses: Exposing and elucidating the ethical issues. *Science and Engineering Ethics.* doi:10.1007/s11948-016-9792-z.

Kaiser, M., E.-M. Forsberg, B. Mepham, K. Millar, E. Thorstensen, and S. Tomkins. 2004. Decision-making frameworks. In *Evaluation of Ethical Bio-Technology Assessment Tools for Agriculture and Food Production. First Interim Report. Ethical Bio-TA Tools (QLG6-CT-2002-02594),* ed. V. Beekman. http://www.ethicaltools.info/content/Interim_Report_Description.pdf.

Koch, T. 2012. *Thieves of virtue: When bioethics stole medicine.* Cambridge: MIT Press.

Montgomery, J. 2016. Bioethics as a governance practice. *Health Care Analysis* 1–21.

Moula, P., and P. Sandin. 2015. Evaluating ethical tools. *Metaphilosophy* 46 (2): 263–279.

Rhodes, R. 2015. Good and not so good medical ethics. *Journal of Medical Ethics* 41 (1): 71–74.

Rip A., and H. te Kulve. 2008. Constructive technology assessment and socio-technical scenarios. In *Presenting Futures, Volume 1 of the series The Yearbook of Nanotechnology in Society,* ed. E. Fisher, C. Selin, and J.M. Wetmore, 49–70. Springer.

Rothman, D.J. 1991. *Strangers at the bedside.* New York: Basic.

Rowe, G., and L.J. Frewer. 2000. Public participation methods: A framework for evaluation. *Science, Technology and Human Values* 25: 3–29.

Seedhouse, D. 2009. *Ethics: The heart of health care.* Chichester, U.K.: Wiley-Blackwell.

Sheehan, M., and M. Dunn. 2013. On the nature and sociology of bioethics. *Health Care Analysis* 21 (1): 54–69.

Stevens, M.L.T. 2003. *Bioethics in America.* Baltimore: Johns Hopkins University Press.

Stirling, A. 2008. "Opening up" and "closing down": Power, participation, and pluralism in the social appraisal of technology. *Science, Technology and Human Values* 33: 262–294.

Sugarman, J., and D.P. Sulmasy. 2010. *Methods in medical ethics.* Washington: Georgetown University Press.

Chapter 3
Ethical Concerns in HCE: The Examples of Cognitive Enhancing Drugs and Noninvasive Brain Stimulation

3.1 Methodology

In the mapping of ethical issues regarding human cognitive enhancement, we used a two-tiered approach in which we first carried out a systematic search, followed by the addition of references found in the literature identified in this search. The first round of search provided a range of topics, but also many new references that we saw fit to include.

In the first search, we used Thompson-Reuter's Web of Science using very open search terms: "ethics" and "cognitive enhancement". We found 87 articles that we scrutinised. We removed entries that were clearly off topic or that were in some sense incomplete, for example, abstracts to conferences and non-peer reviewed articles. 30 papers remained. The majority of papers concerning specific applications focused on pharmaceutical enhancers, with only a few addressing non-invasive brain stimulation techniques. After having read through these articles, we added references that addressed arguments that the first set of articles either supported or tried to refute. In addition, papers addressing non-invasive brain stimulation specifically were added. This left us with a total of 67 papers.

The ethical issues that are discussed in the context of the two applications of interest here fall under two main categories. The first category concerns health issues and includes items such as safety and efficacy. The second category concerns other individual or societal consequences of the use of these forms of cognitive enhancement and includes issues such as fairness and personal achievement, distributive justice and coercion. Another ethical issue that is frequently discussed in the context of noninvasive brain stimulation techniques is the issue of autonomy and consent, particularly with regard to the use of these techniques on children. While authenticity and naturalness are key issues within the overall enhancement debate, many of the papers we reviewed either did not engage with these concepts or considered them as part of another argument. However, given the general importance of this topic to the broader enhancement debate, we include it here.

© The Author(s) 2017
E.-M. Forsberg et al., *Evaluating Ethical Frameworks for the Assessment of Human Cognitive Enhancement Applications*, SpringerBriefs in Ethics, DOI 10.1007/978-3-319-53823-5_3

Privacy is an important issue for some HCE applications (such as smart glasses), but not for the two applications discussed here. Even if privacy is not included here, an ethical framework should be able to address also such an issue. An analysis of the ethics of smart glasses has also identified issues of agency, responsibility, social interaction, and power and ideology (Hofmann et al. 2016).

We will here first present the two applications (or rather, areas of applications) that we focused on in our literature search. Then we will go through the main ethical issues that we identified in the search.

3.2 Cognitive Enhancing Drugs

Cognitive enhancing drugs, also called smart drugs or "nootropics"—from the Greek roots *noo-*, mind and *-tropo*, turn/change (Cakic 2009)—are used to treat cognitive disabilities and improve the quality of life for patients with neuropsychiatric disorders and brain injury (Sahakian and Morein-Zamir 2011). Such drugs are used in treating cognitive impairment in disorders including Alzheimer's disease, schizophrenia and Attention Deficit Hyperactivity Disorder (ADHD) (Sahakian and Morein-Zamir 2011). Many of these same drugs have also been used by healthy individuals in an attempt to gain "better than normal" cognitive ability (Farah et al. 2014; Hall 2004; Racine and Forlini 2009). Enhancing effects of cognition can only be shown on the level of distinct functions such as concentration, alertness, working memory, long-term memory and so on (Schermer et al. 2009).[1] The most commonly used drugs for cognitive enhancement are stimulants, namely Ritalin (methylphenidate) and Adderall (mixed amphetamine salts) and are prescribed primarily for the treatment of ADHD (Greely et al. 2008). A modest degree of memory enhancement is possible with these ADHD medications (Greely et al. 2008). A newer drug, Modafinil—approved for the treatment of fatigue caused by narcolepsy, sleep apnoea and shift-work sleep disorder—has also shown enhancement potential and has been tried on healthy people who need to stay alert and awake when sleep deprived, such as doctors on night duty (Greely et al. 2008).[2] Academics are reported to make use of Modafinil to counteract the effects of jetlag, to enhance productivity and to deal with demanding intellectual challenges (Sahakian and Morein-Zamir 2007). The advantages and disadvantages of the use of pharmaceutical enhancers for both individuals and society have been discussed (Mohamed 2014; Chan and Harris 2006) and will be discussed in more detail here.

---

[1]The "intelligence" trait is too complex and multi-faceted to be enhanced by one single intervention (Schermer et al. 2009).

[2]A variety of practices of medication use for enhancement purposes may exist and develop further in future. These include occasional boosts for special occasions; continuous use to improve performance in high-pressure competitive environments; experimentation for curiosity or fun; substance abuse; and auto-medication of mental problems (Schermer et al. 2009).

Survey numbers indicate that the use of cognitive enhancers by students in the United States is likely to be in the range of 5–15% (Ragan et al. 2013) and such use appears to occur primarily at prestigious universities. Studies investigating students' motivations and reasons for the use of HCE have unearthed a variety of motivations. Reasons for use range from overcoming tiredness and sleepiness (Castaldi et al. 2012), to "getting ahead" and maintaining a high level of academic achievement or, conversely, as a method of "keeping up" or coping with stressful tasks, such as exams (Partridge et al. 2013) and to improve concentration (Mache et al. 2012). There are a few studies which have investigated the prevalence of, attitude to and rationale for the use of cognitive enhancers by university students in Europe, specifically in Switzerland (Ott and Biller-Andorno 2014), Germany (Sattler et al. 2013; Hildt et al. 2014) and in the United Kingdom and Ireland (Singh et al. 2014). A study by Forlini et al. (2015) regarding the prevalence, views and knowledge of a large sample of German students from three different universities has shown that while neuroenhancement is a well-known phenomenon among German students, only 2.2% of their sample of 1026 reported having used a prescription medicine for enhancement. The predominant motivations for use included exams and competitive situations. However, on the whole, students were unenthusiastic and critical of the use of neuroenhancers in an academic context.

Given the ageing population in many countries and the attendant extended lifespan of individuals, it is also highly likely that cognitive-enhancing drugs that can improve memory in healthy elderly people and will be sought after (Hall 2004; Sahakian and Morein-Zamir 2011).

3.3 Noninvasive Brain Stimulation

Noninvasive[3] brain stimulation (NIBS) techniques such as transcranial magnetic stimulation (TMS) and transcranial direct current stimulation (tDCS) are used as investigative tools in cognitive neuroscience and are increasingly being explored as treatments for a variety of neurological and psychiatric conditions (Hamilton et al. 2011). Noninvasive brain stimulation also has the potential to enhance neurological function in cognitive skills, mood and social cognition (Hamilton et al. 2011).

TMS makes use of electromagnetic induction and involves the generation of a rapid time-varying magnetic field in a coil of wire (Farah et al. 2014). When this coil is held to the head of a subject, the magnetic field penetrates the scalp and skull, inducing a small current parallel to the plane of the coil that is sufficient to depolarise neuronal membranes and generate action potentials (ibid). Different TMS paradigms use a variety of pulse frequencies, intensities and stimulation locations to

---

[3]We only discuss noninvasive brain stimulation here, as the use of noninvasive brain stimulation faces much lower hurdles for non-therapeutic use than is the case for invasive technology (Heinrichs 2012).

achieve specific diagnostic, therapeutic and environmental effects (ibid). In transcranial direct current stimulation (tDCS), weak electrical currents, for example 1 mA, are applied for a short duration (approximately 20 min) to the head via electrodes that are placed on the scalp. The currents pass through the skull and alter spontaneous neural activity (Kadosh et al. 2012).

It has been suggested that TMS could be used as a form of cognitive enhancement in the future (De Jongh et al. 2008; STOA 2009). Studies with the explicit objective of inducing cognitive enhancing effects represent only a small fraction of research using noninvasive brain stimulation in healthy participants (Nuffield Council 2013). However, there are many examples in the scientific literature reporting effects ranging from memory, language skills, vision, mathematical ability and reasoning to emotional processing and mood (ibid). While TMS offers greater spatial and temporal resolution than tDCS, tDCS is less expensive, far more portable, very well-tolerated and associated with fewer safety concerns (Hamilton et al. 2011). Indeed, the low cost of buying or building one's own personal tDCS device has garnered much interest within the DIY community (Fitz and Reiner 2013). One only needs a 9 V battery and other inexpensive and easy-to-source electronic parts and basic instructions (Fitz and Reiner 2013; Lapenta et al. 2014). Conversely, the social penetration of TMS as a *product* is likely to remain low given the high costs, while the use of TMS as a *service* might have a moderate level of social penetration (Dubljević 2015).

3.4 Risk and Efficacy

The main ethical argument for HCE is that improved cognitive skills will lead to better lives. Sandberg and Bostrom state that low intelligence "increases the risk for accidents, negative life events, and low income" (2006, 201), while higher intelligence is related to improved health and greater wealth. Increased cognitive capacities will then, according to Sandberg and Bostrom, reduce the likelihood of harm and increase the likelihood of benefits. Their argument rests on the notion that different types of HCE actually increase cognitive functions and that the increase in cognitive functions is not offset by other negative consequences.

Notwithstanding the view above, concern about safety is a major issue in ethical discussion of cognitive enhancement. While safety is a concern for all medications and procedures, our tolerance for risk is smallest when the treatment is purely elective (Farah et al. 2004). In comparison to other comparably elective treatments such as cosmetic surgery, cognitive enhancement involves intervention into a far more complex system (the brain) and an associated greater risk of unanticipated problems (ibid). Moreover, the trade-off between side effects and improvements may be less clear if healthy individuals use pharmaceutical enhancers to improve their mental performance (Hall 2004). Crucially, and in opposition to the argument advanced by Sandberg and Bostrom above, some scholars observe that "more" may not always be "better" in terms of memory or attentiveness as unanticipated problems could arise (Whetstine 2015). Altering the selective process of memory could have associated effects whereby gains in one area may lead to diminishments in another area (Whetstine 2015). For example, some studies have shown that Adderall may increase focus and attention while reducing creativity. Another concern is that "it may not be possible to simply amplify memory or cognition without having profound effects on our identity" (Whetstine 2015, 174).

The medical safety of PCEs varies among substances and side effects relate both to the direct pharmacological effects and broader physical and physiological changes (Maslen et al. 2014a). The risk of dependence has been highlighted (Farah et al. 2014) as the risk to individuals using these medications specifically for cognitive enhancement is unknown. Indeed, a number of commentators have called for data to inform the discussion on cognitive enhancement. Some authors argue for data regarding the safety and efficacy of the use of these drugs in healthy individuals in order to strengthen the empirical foundation of the ethical debate (Boot et al. 2012; Maslen et al. 2014a), while others call for data concerning the attitudes of people regarding cognitive enhancers (Lucke 2012; Nadler and Reiner 2010) in order to inform policy and practice (in general practitioners' offices, schools, universities and workplaces).

Notwithstanding the huge interest in PCE from philosophers and scientists, evidence as to their effectiveness is still inconclusive (Maslen et al. 2014a). While all three classes of medication have been reported to enhance performance in certain laboratory cognitive tasks for at least some normal healthy subjects, the true reliability and size of these effects, and their usefulness for real-world cognitive enhancement have not been definitively established (Farah et al. 2014). Moreover, most PCEs are only effective in the case of decreased conditions such as sleep deprivation. While such uses are non-therapeutic, the conditions in question are more like conditions requiring treatment than to a state of normal functioning (STOA 2009). Difficulties in assessing efficacy may very likely mean that many products with unproven claims will enter the market (Hall 2004). Experience with purported "natural" forms of enhancement such as nutraceuticals, functional foods and dietary supplements demonstrates that protecting consumers from new technologies with doubtful efficacy will be a challenge (ibid). Given these observations, major scholars in the field have argued that the term "cognitive enhancement" itself is debatable in so far as it implies efficacy that has not been established (Forlini et al. 2013; Hall and Lucke 2010; Racine and Forlini 2009).

3.5   Authenticity and Naturalness

The debates involving the topics of authenticity and naturalness highlight disagreement regarding the meaning of these two terms. A key disagreement is where to draw the line between natural and unnatural enhancement. Whereas some enhancements, such a yoga, are seen as natural and unproblematic, other enhancements, such as genetically engineering fertilised eggs, are seen as highly unnatural and problematic. When using the notion of naturalness as an argument against HCE, the main challenge is to defend an understanding of this term that allows one to draw a line between permissible and impermissible applications between these extremes.

Maslen et al. (2014a) distinguish between the concern about authenticity and naturalness. One concern regarding authenticity is of a purely philosophical nature, concerning numerical personal identity (DeGrazia 2005). Other rather philosophical concerns that are also noted as ethical issues are what it is for an individual to become more or less his "real" self, and similar existential issues (The President's Council of Bioethics 2003; Kass 2003).

Kass (2003) offers an understanding of naturalness as "intelligibility in human terms": coffee or alcohol is intelligible to humans in a different way than pills are.

While recognizing the possible difference-reducing potentials of pharmaceutic enhancers, Kass holds that there is "a sense that the 'naturalness' of means matters" (Kass 2003, 22). He focuses on the meaning and the contexts that immerse the different means that humans use in order to strive for a given end. Kass claims that humans cannot understand the meaning of biomedical interventions' effects on the human body and mind in human terms since they are removed from their previous meaning-providing contexts that typically characterize coffee, cigarettes, training, education etc. He holds that "[t]he lack of 'authenticity' sometimes complained of in these discussions is not so much a matter of 'playing false' or of not expressing one's 'true self', as it is a departure from 'genuine,' unmediated, and (in principle) self-transparent human activity" (2003, 23).

Maslen et al. claim that the authenticity questions fundamentally refer to an assumption that human beings are most authentic when they are in a "natural" state (2014a, 5), an assumption they argue against. Rather, they see authenticity as autonomy, where the individual is free to improve themselves. They claim that if pharmacological enhancers "can, for example, help an individual to concentrate better so that he or she can achieve the goals he or she values, this acts in service of authenticity rather than undermines it" (Maslen et al. 2014a, 4). Maslen et al. seem to agree with Kass (2003) that pills that produce completely new cognitive abilities might have a relevant novelty, and as such may be seen as unnatural. They recommend further research into such novel abilities, but still claim that the question of naturalness as non-novelty is ultimately a normative question. They claim that pharmacological HEs at present are "more of the same" and do not change human beings in a novel way.

Sandberg and Bostrom's (2006) core claim is that HCE is a means to increase the likelihood of a good life. The scholars who see authenticity and naturalness as arguments against the desirability of HCE see it as a poor means for a good life (McKibben 2004; Ida 2009; Agar 2014). The Commission of the Bishops' Conferences of the European Community made a statement about the limits of the human condition (Comece 2008, 6). Elliott (2003) and Sandel (2004) also take different strategies in this direction, but at a more general level than simply about pharmacological HEs.

3.6 Fairness and Personal Achievement

Another frequently discussed issue concerns the question as to whether the use of cognitive enhancers—in exams or at work, for example—constitutes cheating, conferring an unfair advantage over others in competitive situations, resulting in an "unlevel playing field" (Cakic 2009; Sahakian and Morein-Zamir 2011).[4] A related ethical issue noted by Maslen et al. (2014a, 8) goes beyond "fairness in competitive contexts to ask whether personal achievement facilitated by PCEs are devalued for this reason". A number of scholars, however, posit arguments which undermine concerns regarding cheating. One argument counters that, given widespread biological and environmental inequalities already in existence, the validity of the level playing field concept can be questioned (Cakic 2009; Dresler et al. 2013). Cakic (2009) makes this argument with respect to genes and the socioeconomic background of one's parents which also have an impact on conferring advantage over others. Another argument introduces the relevance to the debate of the particular neural systems affected by different substances and their disparate effects; in other words, "whether a substance improves creativity or rote learning may matter for some possible conceptions of what constitutes cheating" (Maslen et al. 2014a, 8). Goodman (2010) argues that the use of cognitive enhancing drugs does not cheapen accomplishments achieved under their influence; cognitive enhancement can, rather, be seen as being in line with well-established conceptions of collaborative authorship, in which the locus of praise and blame can be shifted from individual creators to the ultimate products of their efforts. Schermer (2008) highlights the importance of understanding education and other arenas of activity such as sport as "practices" with their own internal goods and standards of excellence, the understanding of which can facilitate the articulation of potential problems of enhancement.

The likelihood of cognitive trade-offs adds another dimension to the cheating debate (Maslen et al. 2014a): evidence suggesting that enhancement in some domains comes at the cost of impairments in other domains challenges the view that achievements enabled by PCE do not involve sufficient personal sacrifice (which is a factor in the judgement of whether something is cheating). Biedermann (2010) mentions the possibility for a future where efforts are seen as unnecessary striving, and technological or pharmaceutical measures take the role of hard work. This resonates with Forlini and Racine's (2009) study on the use of Ritalin in academic settings where all stakeholders agreed that Ritalin was an "easy way out" and that such use connoted dishonesty.

---

[4]Interestingly, while universities have academic codes of conduct that prohibit cheating and plagiarism, they have yet to directly address the use of cognitive enhancers as violations of academic integrity, as they "are regarded in a moral gray zone" (Whetstine 2015, 175).

3.7   Distributive Justice

Society-level debates about PCE-related inequality consider distributive justice and the issue as to whether PCEs will worsen existing socio-economic inequality, particularly if only the wealthy can access them (Biedermann 2010; Maslen et al. 2014a). Proponents of human enhancement counter this argument with two responses. First, they argue that this is more a criticism of existing social hierarchies than a convincing objection to enhancement per se (Hall 2004). Second, they argue that the problem can be overcome by addressing inequities in access to the new technologies (ibid). For example, all forms of enhancement could be made freely available to everyone through public subsidising of costs. Another argument suggests that such distribution of enhancers would contribute to progress in developing countries or among societal groups (such as elderly people with mild cognitive deficiencies or children in areas of poverty) in the developed world, and would, overall, lead to an improvement in the human condition (Nam 2015). If some PCEs are affordable, they could be adopted in disadvantaged populations, much like what happened with the mobile phone in the developing world (Sahakian and Morein-Zamir 2011).

Biederman (2010) notes that it is an open question as to whether HCE should be regarded through the lens of a zero-sum game (see also Buchanan 2008), i.e. there might be collective goods arising from HCE, in addition to collective costs. This issue also touches upon the claimed competitive advantage societies and/or individuals will experience from the use of HCE. Beyer et al. (2014) argue that PCEs will mainly be used by those who are well-off and suggest a taxation on PCEs whereby the governmental income should be earmarked for health-initiatives for the worst off. Dunlop and Savulescu (2014) suggest giving priority of HCE to persons with an IQ in the lowest range (IQ < 75), arguing that the occurrence of such a low IQ level correlates with a range of social ills such as unemployment, underemployment, poverty and chronic welfare dependence. An increase in IQ for these people will have both individual benefits since it reduces the mentioned ills, but also a societal benefit since they will be less dependent on welfare as a collective. Dunlop and Savulescu are not committed to any specific enhancer, but mention modafinil and methylphenidate as the stimulants having the greatest potential. They are also open to more conventional methods, including adding iodine to diets, as this can increase IQ by 10–15 points for those lacking iodine and costs only 2–3 cents per year.

On a more speculative note, Proust warns us of a potential "cognitive arms-race that can only be detrimental to mankind" (2011, 167) in which individuals and states would compete with each other in order to achieve the most encompassing HCE. Proust concludes that a "principle of scientific responsibility would argue for selecting the areas where cognitive and emotional enhancement would be fruitfully enhanced, while banning research in which enhancements are predictably conducive to violence, addiction, irrepressible consumerism, submission, and, in general, to behaviors that violate the agent's dignity or autonomy" (Proust 2011, 167).

Biedermann (2010) claims that the only certain beneficiary of a liberal regime on PCE will be pharmaceutical companies (see also Micoulaud-Franchi et al. 2012).

Another argument concerns whether enhancement interventions might take resources away from more useful medical research targeted at serious diseases that could affect the well-being of the poor majority of the world (Giubilini and Sanyal 2015). Such egalitarian concerns can also work to justify the normative significance of the therapy-enhancement distinction (Giubilini and Sanyal 2015). Thus, it has been argued that, given the limited resources available, therapy has priority over enhancement because making everybody a "normal competitor" is necessary to maintain fair equality of opportunity for different members of society. Other scholars take a different view and question whether it would be ethical to deny healthy individuals a cognitive enhancer that has been shown to be perfectly safe and reliable (Sahakian and Morein-Zamir 2007). This argument relates to the cognitive liberty argument discussed above.

3.8   Coercion

Coercion, either explicitly or implicitly, to take cognitive enhancers comprises another issue that frequently arises in ethical debate. Coercion can be viewed as a "social consequence" of neuroenhancement (Heinz et al. 2012). Such coercion may occur explicitly by, for example, requiring workers to be alert during a night shift or it may occur more implicitly, such as in establishing a competitive environment in which incentives are offered for best performance (Sahakian and Morein-Zamir 2011). Children represent a special case here as they are unable to make their own decisions (Greely et al. 2008; Gaucher et al. 2013). This concern also extends to students who may experience implicit pressure to take PCEs in order to keep up with their peers (Cakic 2009; Biedermann 2010). However, Cakic lists a number of criteria which he argues need to be fulfilled in order for a student to be indirectly coerced into using PCEs. First, enhancers would have to confer significant improvements in performance such that not taking them would leave students at a distinct academic disadvantage relative to their peers. Second, a sufficiently high proportion of the student's peer body would need to be taking enhancers in order to justify the perception that "everybody else is taking them". Finally, the most successful students would need to be using PCEs so as to "perpetuate the presumption that it is either impossible or prohibitively difficult for a drug-free student to attain high grades" (Cakic 2009, 612). Again, given the lack of data regarding the prevalence and use of PCEs in academia, one can only speculate as to whether this situation actually exists in (some) current academic environments.

There is also another element to this coercion argument. Some argue that the approach of banning or restricting the use of neurocognitive enhancement at school or in the workplace is also coercive as it "denies people the freedom to practice a safe means of self-improvement, just to eliminate any negative consequences of the (freely taken) choice not to enhance" (Farah et al. 2004, 423). In other words, banning enhancement technologies altogether would just replace one form of coercive control with another (Hall 2004).[5] In an article studying health workers', students' and parents' views regarding Ritalin and coercion, Forlini and Racine found that their respondents converged on the view that the use of CE is seen at once to be a personal choice and a "result of tremendous social pressures to perform and succeed in very competitive environments" (2009, 166). This means that even though there is a strict separation in much normative ethics between coercion and personal choice, these two elements are easily combined by key stakeholders. Forlini and Racine explain this in their observation that "participants advanced the role of autonomy at the normative level. However, at the descriptive level, social pressures were abundantly illustrated by an overwhelming majority of focus group participants" (2009, 174). The coercive aspect is seen as the need to obtain good grades for future success, while the element of free and personal choice is that the students should be free to be who they want. Forlini and Racine interpret the positions discovered in their focus groups as social pressure that limits the domain for personal choice.

3.9 Ethical Issues Specific to Non-invasive Brain Stimulation

Kadosh et al. (2012) list ethical issues of cognitive enhancement using tDCS differing from those raised by pharmacological interventions. First, the relative inexpense and portability of tDCS means that its use is not limited to laboratories or clinics; we have already mentioned the DIY community that has emerged around this technique and some companies already offer the device for personal use by adults at home. Second, unlike PCEs, tDCS is not ingested into the body. People may perceive a moral difference between "external" enhancements, such as education or computing, and "internal" enhancements, such as drugs that may have worrisome consequences. Kadosh et al. (2012) point out that "[t]he intuition that tDCS is an external intervention may create the misplaced perception that its use is less problematic than more obviously internal enhancements, and thus lower the threshold for premature use" (p. 108). Finally, tDCS can be applied to any cortical brain area, including areas beyond that for which its use may be indicated. In addition, tDCS can have enduring effects. While tDCS has been praised for inducing only transient changes in the brain, studies have reported effects lasting for months. The concern here is that users may bring about long-lasting effects in their underlying neurobiology which may be difficult to reverse (Fitz and Reiner 2013). The possibility of long-lasting effects highlights the importance of the impact of value-laden words such as "non-invasive" on safety (or perceptions of safety):

---

[5]Beyer et al. (2014) hold that all regulation and justification of PCEs must take into account autonomy. They further see PCEs as a tool for potentially improving individual autonomy.

although the electrodes do not penetrate the brain, the electrical current must do so, otherwise it would have no effect on neural function. Thus, tCDS is minimally invasive *in some meaningful sense*. Yet the technically correct descriptor 'non-invasive' carries substantial rhetorical power with regard to safety, an issue that is particularly relevant when considering DIY users (Fitz and Reiner 2013, 2)

A particular issue relevant to the use of tDCS or other forms of NIBS methods in children concerns its possible effect on brain development and the degree to which enhancing some capacities may bring about a deterioration in other capacities (Kadosh et al. 2012). While adult brain stimulation is thought to be reasonably safe when used within defined limits, "known unknowns" regarding the effects and side-effects of stimulation, a lack of clear dosing guidelines and a lack of translational studies from adults to children warrant greater investigation into the use of brain stimulation for children (Davis 2014).[6]

Moreover, while adults are in a position to decide whether a particular effect, for example, enhancing a child's long-term memory, is sufficiently valuable (to them) to justify bringing about a particular impairment, children are not equipped with the capacity or life experience to make such trade-off decisions. Informed consent is a key issue in this regard. The effects of brain stimulation for "enhancement" may have consequences that reach far into a child's future. Maslen et al. (2014b) argue that in order to evaluate the reasons one might have for refusing (and, we might add, allowing) an enhancement, one must be capable of "meaningful temporal projection". This forward-looking capacity is particularly important when making decisions about how to weigh the relative value of different cognitive functions. Crucially, younger children do not possess this capacity (Maslen et al. 2014b). Given an uncertain weighing of benefits, risks and costs, their parents' capacity to determine the child's best interests diminish, and the need to protect the child's (future) autonomy becomes more important (Kadosh et al. 2012).

Proust (2011) views the introduction and use of both invasive and non-invasive brain stimulants as a promising method to provide targeted learning methods to children with different learning needs. She further views different forms of HCE as a suitable means for providing those from poor backgrounds with both cognitive and emotional development support.

3.10 Further Arguments in Favour of Cognitive Enhancement

The arguments presented above are generated by a literature search explicitly focusing on "ethics". Ethics is often called for when there are concerns. Thus, this strategy may potentially leave some positive ethically relevant arguments out. We

---

[6]While Davis discusses brain stimulation in relation to treating neurological disorders in pediatric cases (thus for a therapeutic application), the gaps in knowledge that affect our ability to assess risk in translating brain stimulation procedures to pediatric cases similarly apply to its non-therapeutic use.

saw in the introduction of the two technology cases the kinds of benefits these technologies may yield for the individuals using them. In liberalist capitalist societies there is no need for products to demonstrate benefits other than those demonstrated by the fact that there is a market. Therefore, these benefits are usually not presented as ethical arguments. However, without these benefits—and thus a market for such applications—there would not even be any discussion about HCE. These benefits should therefore be included in an ethical assessment framework. Moreover, the evidence base for these claimed benefits should be assessed, in the same way as the evidence for the risks and costs of the enhancement applications. In addition to the arguments for HCE that have come up in the discussions earlier in this chapter, two more arguments are presented before we move on.

Several scholars argue that it would be a restriction to personal freedom if we do not allow for individual choice vis a vis the shaping of one's nature (Buchanan et al. 2000): if we allow for plastic surgery—also carrying along risk to health—why would we ban the non-therapeutic use of cognitive enhancement technology? If we allow for high altitude training for sportsmen to increase the level of oxygen in the blood, why do we ban doping? And if we applaud the use of meditation techniques to reduce stress, why do we frown upon relaxants that perform the same function? This freedom to choose revolves around the fundamental value of individual's autonomy.

Cognitive enhancement may also be seen as necessary for human kind to tackle our grand challenges, such as climate change: "Non-traditional cognitive enhancement might be able to produce a solution to this problem, by generating more adept scientists who can figure out ways to reverse the effects of carbon, or invent more efficient forms of transport, or more adept economists who can sell alternative energy to brighter politicians" (Fenton 2009, 150). Fenton then argues that the immediacy of climate change calls for a speeding up of testing of non-traditional HCE.

3.11 Summary

Summing up, the literature review has revealed that the main ethical issues related to HCE are:

- Safety
- Efficacy
- Fairness
- Personal achievement
- Distributive justice
- Coercion
- Authenticity/naturalness
- Autonomy and consent

Any ethical assessment tool should be able to address such issues.

With this review of ethically relevant concerns and arguments related to the two key cases of HCE, we are now in a position to present and review potential ethical tools systematizing the ethical issues into decision guidance.

References

Agar, N. 2014. *Truly Human Enhancement. A Philosophical Defense of Limits.* MIT Press.

Beyer, C., C. Staunton, and K. Moodley. 2014. The implications of methylphenidate use by healthy medical students and doctors in South Africa. *BMC Medical Ethics* 15: 20. doi:10.1186/1472-6939-15-20.

Biedermann, F. 2010. Argumente für und wider das Cognitive Enhancement. *Ethik in der Medizin,* 22 (4): 317–329. doi:10.1007/s00481-010-0070-3.

Boot, B.P., B. Partridge, and W. Hall. 2012. Letter to the editor: Better evidence for safety and efficacy is needed before neurologists prescribe drugs for neuroenhancement to healthy people. *Neurocase* 18 (3): 181–184.

Buchanan, A. 2008. Enhancement and the ethics of development. *Kennedy Institute of Ethics Journal* 18 (1): 1–34.

Buchanan, A., D.W. Brock, N. Daniels, & D. Wikler. 2000. *From Chance to Choice: Genetics and Justice.* Cambridge: Cambridge University Press.

Cakic, V. 2009. Smart drugs for cognitive enhancement: Ethical and pragmatic considerations in the era of cosmetic neurology. *Journal of Medical Ethics* 35: 611–615.

Castaldi, S., U. Gelatti, G. Orizio, U. Hartung, A.M. Moreno-Londono, M. Nobile, and P. J. Schulz. 2012. Use of cognitive enhancement medication among Northern Italian University students. *Journal of Addict Medicine* 6 (2): 112–117.

Chan, S., and J. Harris. 2006. Cognitive regeneration or enhancement: The ethical issues. *Regenerative Medicine* 1 (3): 361–366.

COMECE. 2008. Ethical questions raised by nanomedicine. Science and Ethics. *Opinions Elaborated by the Bioethics Reflexion Group* 1: 23–28.

Davis, N.J. 2014. Transcranial stimulation of the developing brain: A plea for extreme caution. *Frontiers in Human Neuroscience* 8: 1–4.

De Jongh, R., I. Bolt, M. Schermer, and B. Olivier. 2008. Botox for the brain: Enhancement of cognition, mood and pro-social behavior and blunting of unwanted memories. *Neuroscience and Biobehavioral Reviews* 32: 760–776.

DeGrazia, D. 2005. *Human identity and bioethics.* Cambridge, New York: Cambridge University Press. http://dx.doi.org/10.1017/CBO9780511614484.

Dresler, M., A. Sandberg, K. Ohla, C. Bublitz, C. Trenado, A. Mroczko-Wąsowicz, S. Kühn, and D. Repantis. 2013. Non-pharmacological cognitive enhancement. *Neuropharmacology* 64: 529–543.

Dubljević, Veljko. 2015. Neurostimulation devices for cognitive enhancement: Towards a comprehensive regulatory framework. *Neuroethics* 8: 115–126.

Dunlop, M. and Savulescu, J. 2014. Distributive justice and cognitive enhancement in lower, normal intelligence. *Monash Bioethics Review* 32 (3): 189–204.

Elliott, C. 2003. *Better than well: American medicine meets the American dream.* New York: Norton.

Farah, M.J., J. Illes, R. Cook-Deegan, H. Gardner, E. Kandel, P. King, E. Parens, B. Sahakian, and P.R. Wolpe. 2004. Neurocognitive enhancement: What can we do and what should we do? *Nature Reviews Neuroscience* 5: 421–425.

Farah, M.J., M.E. Smith, I. Ilieva, and R.H. Hamilton. 2014. Cognitive enhancement. *WIREs Cognitive Science* 5: 95–103.

Fenton, A. 2009. Buddhism and neuroethics: The ethics of pharmaceutical cognitive enhancement. *Developing World Bioethics* 9 (2): 47–56.

Fitz, N.S., and P.B. Reiner. 2013. The challenge of crafting policy for do-it-yourself brain stimulation. *Journal of Medical Ethics* 0: 1–3. doi:10.1136/medethics-2013-101458.

Forlini, C., and E. Racine. 2009. Autonomy and coercion in academic "cognitive enhancement" using methylphenidate: Perspectives of key stakeholders. *Neuroethics* 2 (3): 163–177.

Forlini, C., W. Hall, B. Maxwell, S.M. Outram, S. M., P.B. Reiner, D. Repantis, M. Schermer, and E. Racine. 2013. Navigating the enhancement landscape. Ethical issues in research on cognitive enhancers for healthy individuals. *EMBO Reports* 14 (2): 123–128.

Forlini, C., J. Schildmann, P. Roser, R. Beranek and J. Vollmann. 2015. Knowledge, experiences and views of german university students toward neuroenhancement: An empirical-ethical analysis. *Neuroethics* 8 (2): 83–92.

Gaucher, N., A. Payot, and E. Racine. 2013. Cognitive enhancement in children and adolescents: Is it in their best interests? *Acta Paediatrica* 102: 1118–1124.

Giubilini, A., and S. Sanyal. 2015. The ethics of human enhancement. *Philosophy Compass* 10 (4): 233–243.

Goodman, R. 2010. Cognitive enhancment, cheating and accomplishment. *Kennedy Institute of Ethics Journal* 20 (2): 145–160.

Greely, H., B. Sahakian, J. Harris, R.C. Kessler, M. Gazzaniga, P. Campbell, and M.J. Farah. 2008. Towards responsible use of cognitive-enhancing drugs by the healthy. *Nature* 456: 702–705.

Hall, W. 2004. Feeling 'better than well'. *EMBO Reports* 5 (12): 1105–1109.

Hall, W.D., and J.C. Lucke. 2010. The enhancment use of neuropharmaceuticals: More scepticism and caution needed. *Addiction* 105: 2041–2043.

Hamilton, R., S. Messing, and A. Chatterjee. 2011. Rethinking the thinking cap: Ethics of neural enhancement using noninvasive brain stimulation. *Neurology* 76: 187–193.

Heinrichs, J.-H. 2012. The promises and perils of non-invasive brain stimulation. *International Journal of Law and Psychiatry* 35: 121–129.

Heinz, A., R. Kipke, H. Heimann, and U. Wiesing. 2012. Cognitive neuroenhancement: False assumptions in the ethical debate. *Journal of Medical Ethics* 38 (6): 372–375.

Hildt, E., K. Lieb, and A.G. Franke. 2014. Life context of pharmacological academic performance enhancement among university students—a qualitative approach. *BMC Medical Ethics* 15: 23. doi:http://doi.org/10.1186/1472-6939-15-23.

Hofmann, B., D. Haustein, and L. Landeweerd. 2016. Smart-glasses: Exposing and elucidating the ethical issues. *Science and Engineering Ethics*. doi:10.1007/s11948-016-9792-z.

Ida, R. 2009. *Should We Improve Human Nature? An Interrogation from an Asian Perspective. In Human Enhancment*, eds. J. Savulescu and N. Bostrom. Oxford University Press.

Kadosh, R.C., N. Levy, J. O'Shea, N. Shea, and J. Savulescu. 2012. The neuroethics of non-invasive brain stimulation. *Current Biology* 22 (4): R108–R111.

Kass, L.R. 2003. Ageless bodies, happy souls. *The New Atlantis* 1: 9–28.

Lapenta, O.M., C.A. Valasek, A.R. Brunoni, and P.S. Boggio. 2014. An ethical discussion of the use of transcranial direct current stimulation for cognitive enhancement in healthy individuals: a fictional case study. *Psychology & Neuroscience* 7 (2): 175–180.

Lucke, J.C. 2012. Empirical research on attitudes toward cognitive enhancement is essential to inform policy and practice guidelines. *AJOB Primary Research* 3 (1): 58–60.

Mache, S., P. Eickenhorst, K. Vitzthum, B.F. Klapp, and D.A. Groneberg. 2012. Cognitive-enhancing substance use at German universities: Frequency, reasons and gender differences. *Wiener Medizinische Wochenschrift* 162: 262–271.

Maslen, H., N. Faulmüller, and J. Savulescu. 2014a. Pharmacological cognitive enhancement—how neuroscientific research could advance ethical debate. *Frontiers in Systems Neuroscience* 8: 1–12.

Maslen, H., B.D. Earp, R. Cohen Kadosh, and J. Savulescu. 2014b. Brain stimulation for treament and enhancement in children: An ethical analysis. *Frontiers in Human Neuroscience* 8: 1–5. doi:10.3389/fnhum.2014.00953.

McKibben, B. 2004. Enough. Staying human in an engineered age. *An Owl Book*. New York: Henry Holt and Company.

Micoulaud-Franchi, Jean-A, J. Vion-Dur, and C. Lancon. 2012. Peut-on Prescrire Des Psychostimulants Chez Un Étudiant Sain? *Exemple D'un Cas Clinique*. *Thérapie* 67 (3): 213–221.

Mohamed, A.D. 2014. Neuroethical issues in pharmacological cognitive enhancement. *WIREs Cognitive Science* 5: 533–549.

Nadler, R.C., and P.B. Reiner. 2010. A call for data to inform discussion on cognitive enhancement. *BioSocieties* 5 (4): 481–487.

Nam, J. (2015). Biomedical Enhancements as Justice. *Bioethics* 29 (2): 126–132. https://doi.org/10.1111/bioe.12061

Nuffield Council on Bioethics. 2013. *Novel neurotechnologies: Intervening in the brain*. London: Nuffield Council on Bioethics.

Ott, R. and N. Biller-Andorno. 2014. Neuroenhancement among Swiss Students – A Comparison of Users and Non-Users. *Pharamcopsychiatry* 47: 22–28.

Partridge, B., S. Bell, J. Lucke and W. Hall. 2013. Australian university students' attitudes towards the use of prescription stimulants as cognitive enhancers: Perceived patterns of use, efficacy and safety. *Drug and Alcohol Review* 32: 295–302.

Proust, J. 2011. Cognitive enhancement, human evolution and bioethics. *Journal International De Bioéthique*. 23 (3–4): 149–170.

Racine, E., and C. Forlini. 2009. Expectations regarding cognitive enhancement create substantial challenges. *Journal of Medical Ethics* 35: 469–470.

Ragan, C.I., I. Bard, and I. Singh. 2013. What should we do about student use of cognitive enhancers? An analysis of current evidence. *Neuropharmacology* 64 (2013): 588–593. http://dx.doi.org/10.1016/j.neuropharm.2012.06.016

Sahakian, B.J., and S. Morein-Zamir. 2011. Neuroethical issues in cognitive enhancement. *Journal of Psychopharmacology* 25 (2): 197–204.

Sahakian, B., and S. Morein-Zamir. 2007. Professor's little helper. *Nature* 450 (20): 1157–1159.

Sandberg, A., and N. Bostrom. 2006. Converging cognitive enhancements. *Annals of the New York Academy of Sciences* 1093 (1): 201–227.

Sandel, M. 2004. The case against perfection. *Atlantic Monthly*, 51–62.

Sattler, S., C. Sauer, G. Mehlkop, and P. Graeff. 2013. The rationale for consuming cognitive enhancement drugs in university students and teachers. *PLoS ONE* 8 (7): e68821.

Schermer, M. 2008. On the argument that enhancement is "cheating". *Journal of Medical Ethics* 34: 85–88.

Schermer, M., I. Bolt, R. de Jongh, and B. Olivier. 2009. The Future of psychopharmalogical enhancements: Expectations and policies. *Neuroethics* 2 (2): 75–87.

Science and Technology Options Assessment (STOA). 2009. *Human enhancement study (IP/A/STOA/FWC/2005-28/SC35, 41 and 45)*, ed. C. Coenen, M. Schuijff, M. Smits, P. Klaassen, L. Hennen, M. Rader, and G. Wolbring. https://www.itas.kit.edu/downloads/etag_coua09a.pdf. Accessed October 22, 2015.

Singh, I., I. Bard, and J. Jackson. 2014. Robust resilience and substantial interest: A survey of pharmacological cognitive enhancement among university students in the UK and Ireland. *PLoS ONE* 9 (10): e105969.

The President's Council on Bioethics. 2003. *Beyond Therapy: Biotechnology and the Pursuit of Happiness*. New York: Harper Perennial.

Whetstine, L.M. 2015. Cognitive enhancement: Treating or cheating? *Seminars in Pediatric Neurology* 22: 172–176.

Chapter 4
Frameworks Relevant for Assessing HCE Applications

As there are no specific frameworks proposed for addressing ethical issues at the decision-making level in HCE, we have been forced to search for relevant frameworks in neighbouring fields. It is then immediately relevant to look to biomedicine and biotechnology, as many applications that can be used for HCE purposes are from these fields: pharmaceuticals, surgery based applications and biosensors. Moreover, the field of bioethics provides a series of frameworks with a variety of merits. HCE applications may also be ICT based, indicating that the field of ICT ethics may provide relevant frameworks. In the following we will describe some frameworks from these fields.

We do not include here the large number of technology assessment (TA) approaches that have been developed for societal deliberation on (new) technologies (e.g. Guston and Sarewitz 2002; Fisher et al. 2006; Grin and Grunwald 2000; Rip and te Kulve 2008). Such approaches generally describe procedures for societal deliberation on technology issues. Although obviously closely related to ethical values they should not be called ethical frameworks unless they somehow explicitly refer to ethical values, principles or theories. In this respect, we take a stricter approach to ethics than the Ethical Bio TA Tools project did, which included methods such as risk analysis, Delphi method and the Precautionary Principle; methods that do not in themselves refer to ethics explicitly. However, there are hybrid forms of ethics and TA approaches, like the Ethical Technology Assessment approach developed by Palm and Hansson (2006). These have been included in our review.

We also do not consider theories that include a certain prioritisation of moral values or principles. An example is John Rawls' theory of justice (1971, 1993) or Amartya Sen's Capability Approach (1992, 2009). These approaches provide substantive advice on a decision based on the prioritisation of values justified by the respective authors. Here we assume that an ethical framework that is to be useful for a variety of decision makers in liberal societies should not require the user to subscribe to any particular substantive moral approach.

© The Author(s) 2017
E.-M. Forsberg et al., *Evaluating Ethical Frameworks for the Assessment of Human Cognitive Enhancement Applications*, SpringerBriefs in Ethics, DOI 10.1007/978-3-319-53823-5_4

Finally, we will exclude frameworks that clearly are meant to be used by experts. The ETICA framework for ethical analysis of ICT is a good example (Stahl 2011). Though it provides an illuminating overview of ethical issues, it requires much specification work before it can be used by ordinary policy/decision makers. In the following we will present the ethical frameworks that might be candidates for use in the HCE field. In the subsequent chapter, we will assess their respective pros and cons for the purpose of facilitating an ethical assessment of HCE applications as a basis for decision-making by non-philosophers/ethicists.

4.1 Principle Based Ethics

The clearly most established ethical framework in the biomedical field is Beauchamp and Childress' four principles framework, presented in the classic work *Principles of Biomedical Ethics*, originally published in 1979 and with the, until now, 7th revised edition published in 2013. In this book they present, justify and apply their framework mainly in a clinical medical setting. We will here present it in some detail as several of the other frameworks we will consider build on this.

Beauchamp and Childress acknowledge that there are several kinds of norms (rules, rights, virtues and moral ideals) but claim that principles provide the most general and comprehensive norms. Principles "are general norms that leave considerable room for judgment in many cases. Principles therefore do not function as precise action guides that inform us in each circumstance how to act in the way that more detailed rules and judgments do." (Beauchamp and Childress 2001, 13). Rules are similar to principles, only "more specific in content and more restricted in scope" (Beauchamp and Childress 2001, 13). The four principles that serve as a basis in their approach "derive from considered judgments in the common morality and medical traditions" (Beauchamp and Childress 2001, 23) and are the following:

> (1) respect for autonomy (a norm of respecting the decision-making capacities of autonomous persons), (2) nonmaleficence (a norm of avoiding the causation of harm), (3) beneficence (a group of norms for providing benefits and balancing benefits against risks and costs), and (4) justice (a group of norms for distributing benefits, risks, and costs fairly). (Beauchamp and Childress 2001, 12)

These principles refer to the broadly accepted moral theories of deontology, utilitarianism and contractarianism. The principles are points of departure that can be *specified* into rules and/or *balanced* to solve a problem and they are *prima facie*: "A *prima facie* obligation must be fulfilled *unless* it conflicts on a particular occasion with an equal or stronger obligation. This type of obligation is always binding unless a competing moral obligation overrides or outweighs it in a particular circumstance." (Beauchamp and Childress 2001, 14). If an act is both prima facie right and prima facie wrong a balance must be struck between these principles, by determining the relative weights of all competing prima facie norms.

Specification "is a process of reducing the indeterminateness of abstract norms and providing them with action-guiding content" (Beauchamp and Childress 2001, 16), i.e. giving them a specific scope. In this way concrete cases can be subsumed under the norm. Specification can eliminate an apparent dilemma without either applying or balancing norms, but by finding a norm adequately determinate in content to indicate a solution. The beneficence principle can for instance be specified into the rule that "Doctors should put their patients first". But further specification is required to ensure that this rule does not conflict with other prima facie rules, for instance against deception. A specification that solves this potential conflict could be: "In the information they give insurance companies, doctors should describe the medical situation so that the patient will receive the most benefits, as long as it is not outright deception" (Beauchamp and Childress 2001, 16–17). Still, it is not certain whether any particular specification is the morally best justified, even if it provides a solution to a conflict between norms. So, the authors say that "[s]pecification is an attractive strategy for hard cases of moral conflict as long as the specification can be justified" (Beauchamp and Childress 2001, 17). So specifying principles in itself is never sufficient; one must always also evaluate the specification in a broader light: "Nothing in the model of specification suggests that we can avoid judgments that balance different principles or rules in the very act of specifying them." (Beauchamp and Childress 2001, 17–18). They describe the relation between specification and balancing:

> Principles, rules and rights require *balancing* no less than *specification*. We need both methods because each addresses a dimension of moral principles and rules: *range and scope*, in the case of specification, and *weight or strength*, in the case of balancing. Specification entails a substantive refinement of the range and scope of norms, whereas balancing consists of deliberation and judgment about the relative weights or strength of norms. Balancing is especially important for reaching judgments in individual cases, and specification is especially useful for policy development. (Beauchamp and Childress 2001, 18)

Beauchamp and Childress describe in more detail how balancing should be understood: "The metaphor of larger and smaller weights moving a scale up and down graphically depicts the balancing process, but it may also obscure what happens in the process of balancing by suggesting a purely intuitive or subjective assessment. Justified acts of balancing entail that good reasons be provided, not merely that an agent is intuitively satisfied." (Beauchamp and Childress 2001, 18) Specification and balancing are interconnected, but not in a systematic way: "Balancing does often eventuate in specification, but it need not; and specification often involves balancing, but it also might only add details or fill out the commitments of a principle" (Beauchamp and Childress 2001, 19). However, they warn that in very unique cases, trying to substitute balancing with specification would be pointless, unduly complicated or even perilous.

Beauchamp and Childress provide guidance for balancing in a justified way, in terms of the following conditions:

1. "Better reasons can be offered to act on the overriding norm than on the infringed norm (e.g., if persons have a *right*, their interests generally deserve a special place when balancing those interests against the interests of persons who have no comparable right).
2. The moral objective justifying the infringement must have a realistic prospect of achievement.
3. The infringement is necessary in that no morally preferable alternative actions can be substituted.
4. The infringement selected must be the least possible infringement, commensurate with achieving the primary goal of the action.
5. The agent must seek to minimize any negative effects of the infringement.
6. The agent must act impartially in regard to all affected parties; that is, the agent's decision must not be influenced by morally irrelevant information about any party." (Beauchamp and Childress 2001, 19–20).

They explain that "[t]o the extent these conditions themselves incorporate norms, the norms are also prima facie, not absolute" (Beauchamp and Childress 2001, 19). If these conditions are conjoined with "requirements of coherence [...] they should help us achieve a reasonable measure of protection against purely intuitive or subjective judgments" (Beauchamp and Childress 2001, 21). They admit that in some cases it will not be possible to determine which principle should be overriding, but claim that this is a fact of ethics that no theory can mend.

4.2   The Ethical Matrix

In the biotechnological field, several frameworks were explored in the Ethical Biotechnology Assessment Tools project (see Beekman et al. 2006 and Table 4.1). As already noted, it is the decision-making frameworks that are of greatest interest to us. In the Ethical Bio TA Tools project, thirteen decision guiding frameworks were reviewed, but only two of them were deemed as fitting for further testing and evaluation: the ethical Delphi and the ethical matrix. The ethical Delphi is in our opinion not in itself an ethical tool; rather, it is a general procedure for eliciting

Table 4.1   An ethical matrix for assessment of bovine growth hormone in dairy farming

	Well-being	Autonomy	Fairness
Dairy farmers	Satisfactory income and working conditions	Managerial freedom of action	Fair trade laws and practices
Consumers	Food safety and acceptability Quality of life	Democratic, informed choice, e.g. of food	Availability of affordable food
Dairy cows	Animal welfare	Behavioural freedom	Intrinsic value
The biota	Conservation	Biodiversity	Sustainability

expert views that can also be used in an ethics context. This holds also for other general approaches, such as stakeholder analysis. We will therefore not include the ethical Delphi here, but we will discuss the ethical matrix.

The ethical matrix method was developed in the early 1990s by Professor Ben Mepham at the Centre for Applied Bioethics (CAB) at the University of Nottingham for ethical assessment of animal biotechnologies, for instance issues related to technology development in the dairy sector and for evaluation of genetically modified foods. According to Mepham "[t]he aim was to establish a methodology that was versatile in terms of subject matter, user group and form of user engagement" (2004, 271). The aim of the ethical matrix method is to "facilitate, but not to determine, ethical decision-making, and in 'committee use' to identify areas of agreement and disagreement by promoting transparency" (Mepham 2004, 272).

An example of an ethical matrix is the following one, developed for the ethical assessment of bovine growth hormone in dairy cows (Mepham 2005).

The matrix specifies the general principles (well-being, autonomy and fairness) related to each affected party in a certain situation, resulting in an overview of relevant ethical values that need to be considered when introducing such new technologies. This could be used as a starting point for ethical analysis, discussion and ultimately decision making. It is inspired by Beauchamp and Childress and requires the same kind of balancing to yield conclusions, but the principles are slightly modified. In Mepham's view, the difference between beneficence and non-maleficence was not as acute in animal biotechnologies and could therefore be combined into a principle of well-being.

The ethical matrix has been widely applied in the context of biotechnology and in other fields, such as radiation protection, natural management, etc. (Kaiser and Forsberg 2001; Forsberg 2007; Oughton et al. 2004; Jensen et al. 2011; Cotton 2009) and would prima facie appear as a candidate for an ethical framework also in the HCE field.

4.3   Anticipatory Ethics for Emerging Technologies

Brey (2012) presents an ethical framework that is especially targeted to emerging science and technologies. The defining feature of such technologies is that they are not yet in wide societal use so there exist high uncertainty about their consequences and moral implications. According to Brey there is therefore a need to apply anticipatory methods embedded in ethics assessment; and his anticipatory ethics for emerging technologies (ATE) is a proposal for such an approach. Brey distinguishes between several stages in technology development. The first stage, the Research stage (R) is characterised by large uncertainties. At the Development stage (D) there is slightly less uncertainty, but still uncertainties abound. ATE operates at the R and D stage and includes forecasting as a central element.

ATE distinguishes three levels of ethical analysis: the technology, the artifact, and the application level. Various objects of ethical analysis are defined at each of these levels. The *technology level* is the level at which a particular technology is defined, independent of any artifacts or applications that may result from it. An *artifact* refers to a physical configuration that, when operated in the correct manner and environment, produces a desired result. An *application* is defined as the concrete use of a technology artifact or procedure for a particular purpose or in a particular context, or a specific configuration of an artifact to enable it to be used in a particular way (Brey 2012, 8).

Ethical analysis at the *technology level* centres in on the general features of the technology, particular subclasses of it, or techniques within it. It then considers general ethical issues associated with these features: "These are either ethical issues inherent to the character of the technology, issues that pertain to consequences that are likely to manifest themselves in any or nearly any artefact or application of the technology, or issues pertaining to risks that the technology will result in artifacts or applications that are morally problematic" (Brey 2012, 8). Ethical analysis at the *artifact level* focuses on types of artifacts and processes that have resulted, or are likely to result, from a particular technology and the associated features that present moral issues. Such moral issues may arise as a result of the inherent character of the artifact, because the artifact has particular unavoidable consequences in many of its uses, or because certain potential applications of the artifact are so risky or morally controversial that reflection on the ethical justification of manufacture is necessary. Ethical analysis at the *application level* is concerned with particular ways of using an artifact or procedure, or on particular ways of configuring it for use. Ethical issues at the application level fall into three groups. The first group consists of moral issues relating to the intended use of the artifact. Such issues concern the morality of the particular purposes for which an artifact or procedure may be used. A second group is made up of moral issues concerning side-effects or unintended consequences for users. These include consequences that arise in certain uses, in certain contexts of use, or for certain users groups. A third group consists of moral issues concerning the rights and interests of non-user stakeholders who may be affected by a particular use of an artifact. To sum up, at the technology level, fundamental ethical issues concerning the technology are investigated, while more specific and contingent issues are studied at the artifact and application levels.

Knowledge of these objects of analysis can be gained through different forecasting methods, including the use of existing forecasting studies, expert panels and surveys, and self-performed futures studies. Finally, ethical analysis is performed at two initial stages, i.e. the identification and evaluation stages. During the identification stage, moral values and principles are operationalised and cross-referenced with technology descriptions derived from the forecasting stage. The values and issues are derived form an ethical checklist, in addition to identification of ethical issues in the technology ethics literature and bottom-up ethical analyses of the various artefacts and applications. The ethical checklist is structured around four categories of ethical principles that are widely recognised in ethics, i.e. those relating to the prevention of harms, the protection of rights, the pursuit of justice, and the promotion of well-being and the common good (see Table 4.2). The potential importance of ethical issues identified is evaluated during the evaluation

Table 4.2 Brey's ethical checklist

Main category	Sub-category	Specifics
Harms and risks		
	Health and bodily harm	
	Pain and suffering	
	Psychological harm	
	Harm to human capabilities	
	Environmental harm	
	Harms to society	
Rights		
	Freedom	
		Freedom of movement
		Freedom of speech and expression
		Freedom of assembly
	Autonomy	
		Ability to think one's own thoughts and form one's own opinions
		Ability to make one's own choices
		Responsibility and accountability
		Informed consent
	Human dignity	
	Privacy	
		Information privacy
		Bodily privacy
		Relational privacy
	Property	
		Right to property
		Intellectual property rights
	Other basic human rights as specified in human rights declarations (e.g., to life, to have a fair trial, to vote, to receive an education, to pursue happiness, to seek asylum, to engage in peaceful protest, to practice one's religion, to work for anyone, to have a family, etc.)	
	Animal rights and animal welfare	
Justice (distributive)		
	Just distribution of primary goods, capabilities, risks and hazards	
	Nondiscrimination and equal treatment relative to age, gender, sexual orientation, social class, race, ethnicity, religion, disability, etc.	
	North–south justice	
	Intergenerational justice	
	Social inclusion	

(continued)

Table 4.2 (continued)

Main category	Sub-category	Specifics
Well-being and the common good		
		Supportive of happiness, health, knowledge, wisdom, virtue, friendship, trust, achievement, desire-fulfillment, and transcendent meaning
		Supportive of vital social institutions and structures
		Supportive of democracy and democratic institutions
		Supportive of culture and cultural diversity

stage and these issues are elaborated. Evaluations may subsequently be used for improving technology development or for enhanced governance of technology.

4.4 Ethical Technology Assessment

Ethical Technology Assessment (ethical TA) refers to an ethical assessment framework for new and emerging technologies that can be used to explore the soft impacts of such technologies. The idea behind Ethical TA is to render conflicts and differences in opinion on new technologies more explicit rather than evening them out (Palm and Hansson 2006). Ethical TA aims to identify the ethical aspects of an emerging technology to inform and steer design processes to avoid, ex ante, the emergence of ethical controversies. Reaching consensus should, however, not be the pre-set goal of this approach. It focuses on rendering accessible negative aspects of new and emerging technologies on topics such as human relations, values and identities through the discussion of specific scenarios. In an ethical TA the following ethical check list is operationalised:

1. Dissemination and use of information
2. Control, influence and power
3. Impact on social contact patterns
4. Privacy
5. Sustainability
6. Human reproduction
7. Gender, minorities and justice
8. International relations
9. Impact on human values.

In operationalising this checklist, ethical TA aims to further the science-society dialogue and thus enable public participation with technology developers as well as with political decision-makers.

4.5 Ethical Constructive Technology Assessment

Kiran et al. (2015) continue the effort initiated by ethical TA (see above): to fill in the gaps left by regular TA. They believe one should proceed beyond evaluating a list of pre-defined ethical issues; what can be called a 'checklist approach'. In their paper they define principles for an ethical-constructive technology assessment approach (eCTA) that builds onto the philosophy of technology and Science and Technology Studies (STS). Kiran et al. criticise a supposed gap between man and technology in many TA approaches and in the tradition of STS the authors continue by claiming that technologies should be studied as mediators of the relation humans have with their environment. They do so on the basis of the postphenomenological approach of Idhe (1983, 1990) who defines our relation to technology fourfold: through the way it embodies us, the way it provides us with ways to interpret the world, through perceiving technology as another subject and through creating a backdrop to our lives. On this basis they develop a framework to assess the way in which novel technologies also affect the microprocesses in our daily lives. Rather than assessment, the authors suggest a technology "accompaniment" since our morals and values are continuously shaped by our technologies (Kiran et al. 2015, 16). In their opinion, both design practices and the world in which these will ultimately land, will need to be taken into account. This can only be done when design practices "incorporate openness to situatedness, alternative lifeworlds and changing moral routines" (Kiran et al. 2015, 16). According to Kiran et al. one should take into account the way in which technologies shape subjects as well as the way in which they demand a responsible uptake by subjects. Rather than providing an approach for assessment of technologies, they provide for an account of how technology, society and ethics are co-constituted. Thus, technology, society and ethics coevolve and all three should be taken into account simultaneously.

4.6 Ethical Impact Assessment

Building partly on the work in the Ethical Bio TA Tools project, Wright (2011) proposes a framework for an ethical impact assessment which can be performed in regard to any policy, service, project or programme involving information technology. His ethical impact assessment approach offers a means of ensuring that ethical implications are adequately investigated by stakeholders prior to the deployment of a new technology or project in order that mitigating measures can be taken as necessary. An EIA is a process that comprises a number of steps. The first steps include (1) determining whether an EIA is necessary; (2) identifying the EIA team and setting the team's terms of reference, budget and timeframe; (3) preparing an EIA plan; (4) describing the proposed project to be assessed; (5) identifying

Table 4.3 Some main categories and questions from EIA, excluding the privacy issues related more specifically to ICT issues

Respect for autonomy (right to liberty)
Does the technology or project curtail a person's right to liberty and security in any way? If so, what measures could be taken to avoid such curtailment?
Does the project recognise and respect the right of persons with disabilities to benefit from measures designed to ensure their independence, social and occupational integration and participation in the life of the community?
Will the project use a technology to constrain a person or curtail their freedom of movement or association? If so, what is the justification?
Does the person have a meaningful choice, i.e., are some alternatives so costly that they are not really viable alternatives? If not, what could be done to provide real choice?
Dignity
Will the technology or project be developed and implemented in a way that recognises and respects the right of citizens to lead a life of dignity and independence and to participate in social and cultural life? If not, what changes can be made?
Is such a recognition explicitly articulated in statements to those involved in or affected by the project?
Does the technology compromise or violate human dignity? For example, in the instance of body scanners, can citizens decline to be scanned or, if not, what measures can be put in place to minimise or avoid comprising their dignity?
Does the project require citizens to use a technology that marks them in some way as cognitively or physically disabled? If so, can the technology be designed in a way so that it does not make them stand out in a crowd?
Does the project or service or application involve implants? If so, does it accord with the opinion of the European Group on Ethics (EGE)?
Informed consent
Will the project obtain the free and informed consent of those persons to be involved in or affected by the project? If not, why not?
Will the person be informed of the nature, significance, implications and risks of the project or technology?
Will such consent be evidenced in writing, dated and signed, or otherwise marked, by that person so as to indicate his consent?
If the person is unable to sign or to mark a document so as to indicate his consent, can his consent be given orally in the presence of at least one witness and recorded in writing?
Does the consent outline the use for which data are to be collected, how the data are to be collected, instructions on how to obtain a copy of the data, a description of the mechanism to correct any erroneous data, and details of who will have access to the data?
If the individual is not able to give informed consent (because, for example, the person suffers from dementia) to participate in a project or to use of a technology, will the project representatives consult with close relatives, a guardian with powers over the person's welfare or professional carers? Will written consent be obtained from the patient's legal representative and his doctor?
Will the person have an interview with a project representative in which he will be informed of the objectives, risks and inconveniences of the project or research activity and the conditions under which the project is to be conducted?

(continued)

Table 4.3 (continued)

Will the person be informed of his right to withdraw from the project or trial at any time, without being subject to any resulting detriment or the foreseeable consequences of declining to participate or withdrawing?

Will the project ensure that persons involved in the project give their informed consent, not only in relation to the aims of the project, but also in relation to the process of the research, i.e., how data will be collected and by whom, where it will be collected, and what happens to the results?

Are persons involved in or affected by the project able to withdraw from the project and to withdraw their data at any time right up until publication?

Does the project or service collect information from children? How are their rights protected?

Is consent given truly voluntary? For example, does the person need to give consent in order to get a service to which there is no alternative?

Does the person have to deliberately and consciously opt out in order not to receive the "service"?

Non-maleficence

Safety

Is there any risk that the technology or project may cause any physical or psychological harm to consumers? If so, what measures can be adopted to avoid or mitigate the risk?

Have any independent studies already been carried out or, if not, are any planned which will address the safety of the technology or service or trials? If so, will they be made public?

To what extent is scientific or other objective evidence used in making decisions about specific products, processes or trials?

Does the technology or project affect consumer protection?

Will the project take any measures to ensure that persons involved in or affected by the project will be protected from harm in the sense that they will not be exposed to any risks other than those they might meet in normal everyday life?

Can the information generated by the project be used in such a way as to cause unwarranted harm or disadvantage to a person or a group?

Does the project comply with the spirit of consumer legislation (e.g., Directive 93/13 on unfair terms in consumer contracts, Directive 97/7 on consumer protection in respect of distance contracts and the Directive on liability for defective products (85/374/EEC))?

Social solidarity, inclusion and exclusion

Has the project taken any steps to reach out to the excluded (i.e., those excluded from use of the Internet)? If not, what steps (if any) could be taken?

Does the project or policy have any effects on the inclusion or exclusion of any groups?

Are there offline alternatives to online services?

Is there a wide range of perspectives and expertise involved in decision-making for the project?

How many and what kinds of opportunities do stakeholders and citizens have to bring up value concerns?

Isolation and substitution of human contact

Will the project use a technology which could replace or substitute for human contact? What will be the impact on those affected?

Is there a risk that a technology or service may lead to greater social isolation of individuals? If so, what measures could be adopted to avoid that?

Is there a risk that use of the technology will be seen as stigmatising, e.g., in distinguishing the user from other people?

(continued)

Table 4.3 (continued)

Discrimination and social sorting
Does the project or service use profiling technologies?
Does the project or service facilitate social sorting?
Could the project be perceived as discriminating against any groups? If so, what measures could be taken to ensure this does not happen?
Will some groups have to pay more for certain services (e.g., insurance) than other groups?
Beneficence
Will the project provide a benefit to individuals? If so, how will individuals benefit from the project (or use of the technology or service)?
Who benefits from the project and in what way?
Will the project improve personal safety, increase dignity, independence or a sense of freedom?
Does the project serve broad community goals and/or values or only the goals of the data collector? What are these, and how are they served?
Are there alternative, less privacy intrusive or less costly means of achieving the objectives of the project?
What are the consequences of not proceeding with development of the project?
Does the project or technology or service facilitate the self-expression of users?
Universal service
Will the project or service be made available to all citizens? When and how will this be done?
Will training be provided to those who do not (yet) have computer skills or knowledge of the Internet? Who should provide the training and under what conditions?
Will the service cost the same for users who live in remote or rural areas as for users who live in urban areas? How should a cost differential be paid?
Accessibility
Does the new technology or service or application expect a certain level of knowledge of computers and the Internet that some people may not have?
Could the technology or service be designed in a way that makes it accessible and easy to use for more people, e.g., senior citizens and/or citizens with disabilities?
Are some services being transferred to the Internet only, so that a service is effectively no longer available to people who do not (know how to) use computers or the Internet? What alternatives exist for such people?
Values sensitive design
Is the project or technology or service being designed taking into account values such as human well being, dignity, justice, welfare, human rights, trust, autonomy and privacy?
Have the technologists and engineers discussed their project with ethicists and other experts from the social sciences to ensure value sensitive design?
Does the new technology, service or application empower users?
Sustainability
Is the project, technology or service economically or socially sustainable? If not, and if the technology or service or project appears to offer benefits, what could be done to make it sustainable?
Should a service provided by means of a research project continue once the research funding comes to an end?
Does the technology have obsolescence built in? If so, can it be justified?

(continued)

Table 4.3 (continued)

Has the project manager or technology developer discussed their products with environmentalists with a view to determining how their products can be recycled or how their products can be designed to minimize impact on the environment?

Justice

Has the project identified all vulnerable groups that may be affected by its undertaking?
Is the project equitable in its treatment of all groups in society? If not, how could it be made more equitable?
Does the project confer benefits on some groups but not on others? If so, how is it justified in doing so?
Do some groups have to pay more than other groups for the same service?
Is there a fair and just system for addressing project or technology failures with appropriate compensation to affected stakeholders?

Equality and fairness (social justice)

Will the service or technology be made widely available or will it be restricted to only the wealthy, powerful or technologically sophisticated?
Does the project or policy apply to all people or only to those less powerful or unable to resist?
If there are means of resisting the provision of personal information, are these means equally available or are they restricted to the most privileged?
Are there negative effects on those beyond the person involved in the project or trials and, if so, can they be adequately mediated?
If persons are treated differently, is there a rationale for differential applications, which is clear and justifiable?
Will any information gained be used in a way that could cause harm or disadvantage to the person to whom it pertains? For example, could an insurance company use the information to increase the premiums charged or to refuse cover?

stakeholders; and (6) consulting with stakeholders and analysing the ethical impacts. Remaining steps include checking that the project or technology development complies with legislation, preparing and publishing the EIA report and implementing the recommendations.

The EIA framework is partly based on Beauchamp and Childress's four principles, along with a separate part on privacy and data protection. More specific values or issues, explanations and questions for consideration are included for each principle. See Table 4.3 for an overview of values and issues.

As an approach to ethics assessment, EIA is aimed at policy makers, technology developers, and other stakeholders including civil society stakeholders, academics and the media. The person who is carrying out the EIA should be responsible for the conduct of an EIA but may involve additional expertise on the EIA team.

References

Beauchamp, T.L., and J.F. Childress. 2001. *Principles of biomedical ethics*, 5th ed. New York: Oxford University Press Inc.

Beekman, V., E. de Bakker, H. Baranzke, O. Baune, M. Deblonde, E.-M. Forsberg, R. de Graaff, H.-W. Ingensiep, J. Lassen, B. Mepham, A.P. Nielsen, S. Tomkins, E. Thorstensen, K. Millar, B. Skorupinski, F. Brom, M. Kaiser, and P. Sandøe. 2006. *Ethical bio-technology assessment tools for agriculture and food production: Final report ethical bio-TA tools (QLG6-CT-2002-02594)*. Agricultural Economics Research Institute (LEI). https://www.academia.edu/2838074/Ethical_bio-technology_assessment_tools_for_agriculture_and_food_production. Accessed September 1, 2015.

Brey, P.A.E. 2012. Anticipatory ethics for emerging technologies. *Nanoethics* 6 (1): 1–13.

Cotton, M. 2009. Evaluating the 'ethical matrix' as a radioactive waste management deliberative decision-support tool. *Environmental Values* 18: 153–176.

Fisher, E., R.L. Mahajan, and C. Mitcham. 2006. Midstream modulation of technology: Governance from within. *Bulletin of Science, Technology & Society* 26 (6): 485–496.

Forsberg, E.M. 2007. *A deliberative ethical matrix method: Justification of moral advice on genetic engineering in food production*. (Dr. Art). Faculty of Humanities, University of Oslo, Oslo.

Grin, J. and A. Grunwald (eds.). 2000. Vision assessment: shaping technology in 21st century society—Towards a repertoire for technology assessment. *Wissenschaftsethik und Technikfolgenbeurteilung* 4.

Guston, D.H., and D. Sarewitz. 2002. Real-time technology assessment. *Technology in Society* 24: 93–109.

Ihde, D. 1983. *Existential Technics*. Albany: State University of New York Press.

Ihde, D. 1990. *Technology and the Lifeworld. From Garden to Earth*. Bloomington: Indiana University Press.

Jensen, K.K., E.-M. Forsberg, C. Gamborg, K. Millar, and P. Sandøe. 2011. Facilitating ethical reflection among scientists using the ethical matrix as a tool. *Science and Engineering Ethics* 17 (3): 425–445.

Kaiser, M., and E.-M. Forsberg. 2001. Assessing fisheries—Using an ethical matrix in a participatory process. *The Journal of Agricultural and Environmental Ethics* 14: 191–200.

Kiran, A.H., N. Oudshoorn, and P.-P. Verbeek. 2015. Beyond checklists: Toward an ethical-constructive technology assessment. *Journal of Responsible Innovation* 2 (1): 5–19.

Mepham, T. B. 2004. A decade of the ethical matrix: A response to criticisms. In *Science, Ethics & Society*. 5th Congress of the European Society for Agricultural and Food Ethics. Preprints, ed. J.D. Tavernier, and S. Aerts.

Mepham, T.B. 2005. *Bioethics. An Introduction for the Biosciences*. Oxford: Oxford University Press.

Oughton, D., E-M. Forsberg, I. Bay, M. Kaiser, and B. Howard. 2004. An ethical dimension to sustainable restoration and long-term management of contaminated areas. *Journal of Environmental Radioactivity* 74 (1–3): 171–183.

Palm, E., and S.O. Hansson. 2006. The case for ethical technology assessment (eTA). *Technological Forecasting and Social Change* 73: 543–558.

Rawls, J. 1971. *A theory of justice*. Cambridge, MA: Belknap Press of Harvard University Press.

Rawls, J. 1993. *Political liberalism*. New York: Columbia University Press.

Rip, A., and H. te Kulve. 2008. Constructive technology assessment and socio-technical scenarios. In *Presenting Futures, Volume 1 of the series The Yearbook of Nanotehnology in Society*, ed. E. Fisher, C. Selin, and J.M. Wetmore, 49–70. Berlin: Springer.

Sen, A. 1992. *Inequality reexamined*. Oxford: Clarendon Press.

Sen, A. 2009. *The idea of justice*. Cambridge, MA: Belknap Press of Harvard University Press.

Stahl, B.C. 2011. IT for a better future: how to integrate ethics, politics and innovation. *Journal of Information, Communication & Ethics in Society* 9 (3), pp. 140–156.

Wright, D. 2011. A framework for the ethical impact assessment of information technology. *Ethics and Information Technology* 13 (3): 199–226.

Chapter 5
The Adequacy of the Frameworks for Ethical Assessment of HCE Applications

In the previous section, we have seen that there are several ethical frameworks that can be used to assess HCE applications and in Chap. 3 we reviewed the general ethical issues raised by HCE applications. We are now in a position to discuss which of the six above mentioned frameworks seem to fit best for assessing HCE applications. In this section, we will evaluate the frameworks based on their ability to incorporate the most important HCE-related ethical values and concerns (comprehensiveness), their ability to facilitate transparency and their user-friendliness. As explained above, there is a need for a user-friendly approach that will guide non-expert users in considering the most important ethical aspects of specific HCE applications.

5.1 Principle Based Ethics

With regard to comprehensiveness, most of the ethical issues identified above can be sorted under the four principles of Beauchamp and Childress:

- Beneficence: Efficacy
- Non-maleficence: Safety
- Autonomy: Personal achievement, coercion, authenticity
- Fairness: Fairness and distributive justice

In this sense, this approach seems to fit nicely into the HCE context. However, a problem with the four principles approach is that it is directed mainly towards health care professionals, such as medical doctors; a doctor is to do good and avoid doing harm. But in the HCE context there is not necessarily a professional administering the enhancement application.

Another problem with the four principles approach is that it cannot properly deal with the notion of naturalness, though the significance of this problem is up for discussion. Firstly, it is not really clear what the concept of naturalness signifies;

© The Author(s) 2017
E.-M. Forsberg et al., *Evaluating Ethical Frameworks for the Assessment of Human Cognitive Enhancement Applications*, SpringerBriefs in Ethics, DOI 10.1007/978-3-319-53823-5_5

and secondly, it might in fact be more of an aesthetic, than an ethical issue. However, sometimes naturalness is clearly meant as an ethical concept referring to the inherent dignity of man or mankind (Kass 2003). This is a kind of concern that is usually not highlighted in the four principles approach.

One might, however, use the four principles approach and adjust it so that autonomy is replaced with dignity (where the concept of respect for dignity includes the concept of respect for autonomy). This would allow for the inclusion of the issue of naturalness related to human dignity and would thus increase the comprehensiveness of the approach.

One might also reframe the four principles approach to not target a specific professional's responsibility for considering these principles, but to see them more as general values. This would also expand the scope of the approach.

With regard to transparency, Beauchamp and Childress do not offer a clear structure to enhance transparency, leaving it up to the user to transparently report their specification and balancing work. Moreover, despite rules for principles infringement, it is not clear how action guiding principlism is.

Finally, with regard to user friendliness, the widespread use of this approach is a solid testament to this. However, it does require quite a lot of technical and ethical competence to specify and balance the principles.

All in all, the four principles approach (principle based ethics) might be a good starting point for an ethical decision making framework in HCE, but seems to require some further development before it is fully useful. The ethical matrix and the EIA appears to provide this.

5.2 The Ethical Matrix

As described above, the Ethical Bio-TA Tools project identified and described 13 different decision-making frameworks and evaluated these according to user utility, participant satisfaction and whether the tools "capture those arguments, values and principles that various ethical traditions and theories would bring to the fore when dealing with issues of that kind" (Beekman et al. 2006, 20–21). From this review, the ethical matrix emerged as a sound framework (Kaiser et al. 2007, 70).

The ethical matrix approach appears to have an advantage over the four principles approach by its ability to distinguish impacts for the different affected parties. However, this also makes this approach a bit more complex. Whether this complexity makes it more or less user-friendly might be debated and might differ according to user preferences.

The workshop phase with practitioners in Ethical Bio-TA Tools demonstrated that the ethical matrix was not an easy tool to use, but it provided a necessary structure for debates, made evaluations of ethical values more transparent, and opened up for a change of mind on ethical issues in biotechnology (Kaiser et al. 2007). Cotton (2014) seems to have a different view of the ethical matrix and highlights its simplified structure, which "aids simplification and structuring of ethical discussions but also limits opportunities for creative problem solving outside of the matrix's pre-defined principle and stakeholder categories" (73). Jensen et al. (2011) report feedback from workshop participants that clearly support the ethical matrix as a helpful tool (see p. 440).

Cotton also calls for further tools to justify the bottom-up principle and stakeholder selection, and requests an expansion of principles and stakeholders in order to ease potential bias. Cotton's criticism, however, does not hold for the version of the ethical matrix method advanced by Kaiser and Forsberg (2001), as here the content of the matrix is not predetermined. In this version, the matrix must be adjusted in a process of reflective equilibrium so that the structure of principles and stakeholders in fact incorporates all values that are discussed in the field (Forsberg 2007). This will ensure comprehensiveness. We have developed an ethical matrix that is able to incorporate the identified HCE issues (see Table 5.1).

If the ethical matrix approach is to be used as an ethical framework for HCE applications, it is clear that well-being needs to be split into separate columns for positive and negative well-being, otherwise the enhancement dimensions are not well captured. This has been done already in a report for the World Organisation for Animal Health (see Kaiser 2005), and does not amount to a theoretical or practical problem.

In the matrix above, the main four stakeholders for HCE are the user, the non-users, society at large and future generations. However, specific HCE applications might also involve other affected parties (for instance the family of the user).[1] The first step in using an ethical matrix is to adapt it to the specific issue to be considered. This can be done by an individual user, or (if the user has the resources) in a participatory process with interested parties. The higher number of affected parties, the more useful will a matrix structure likely be.

With such a matrix developed for HCE, concrete applications can be assessed on each of the values in the matrix, and a so-called consequence matrix can be filled in portraying the specific ethically relevant consequences of the application. This includes gathering evidence of consequences of use (including, potentially, noting that there are no foreseen positive or negative impact on certain values) and assessing the quality of this evidence. This can then be the basis for systematic and transparent balancing and decision making. For instance, even if a HCE application is seen as unnatural, the amount of benefits it may provide may outbalance this concern. Or if an application is characterised by large uncertainties, vague benefits may not be sufficient for justifying its use. There is no hard algorithm for such balancing, but the matrix structure will make the trade-offs appear much more clearly. The matrix structure is helpful for systematically considering all ethically relevant consequences of a technology choice per affected party.

---

[1]When used in the food production context, producers are also often included as an interested party, as farmers or fishermen often are considered as societal groups that need a certain protection. In the HCE context, it is not equally clear that producers of HCE applications should be considered a group with the same status, so this group is left out of the current example.

Table 5.1 Proposed ethical matrix for HCE

	Increase of well-being	Avoidance of decrease of well-being	Dignity	Justice
The user	Efficacy in providing benefits Increase in quality of life	Safety No reduction in quality of life	Respect for authenticity, naturalness, personal achievement, privacy, and autonomy and consent Avoidance of coercion	Fair access to enhancements Fair access to societal goods
Non-users	Positive effect on their well-being because of others' use of HCE	No negative effect on their well-being because of others' use of HCE	Respect for their privacy and right to choose non-HCE	Fair access to societal goods
The society	Progress in society/improvement of humanity	No societal unintended side effects of the use of HCE	Safeguarding of the room for societal decision making, and not only market forces	Avoiding increased social differences
Future generations	Care for their well-being	No activities that threaten their health or living conditions Precaution	Not diminishing their scope of choice	The conservation of the environment and resources so that future generation will have equal opportunities as we do

In sum, the ethical matrix is more specific than Beauchamp and Childress' approach (and therefore more transparent) and could be a candidate for an ethical decision making framework for HCE.

5.3   Anticipatory Ethics for Emerging Technologies

Brey's three levels of ethical analysis—the technology, artefact and application level—are useful for concretely identifying the focus of ethical analysis within a potentially diffuse HCE discussion. In our context, the application level would seem to be the most relevant. The ATE framework is a practical, transparent and systematic approach that emphasises the importance of knowledge and evidence gathering about the object in question. As already noted, the development of an evidence base for HCE is crucial for ethical deliberation. However, use of the ATE framework requires a significant degree of expert input and resources in terms of the use of various forecasting methods in order to gain knowledge of the objects and the two stages of ethical analysis, namely identification and evaluation. This implies that users would have to have expert resources available to them. Thus, while the ATE is both of high quality, transparent and comprehensive, it is more problematic with regard to our criterion of user-friendliness. Only when expert evaluations have been done, they may be used by non-experts in order to facilitate policy or other forms of decision-making.

5.4   Ethical Technology Assessment

Ethical Technology Assessment is comprehensive, but abstract. The checklist may potentially include all the identified topics, but also much more. Even if the checklist is short and straightforward, it remains highly challenging for a decision maker without expertise to fill all the general points with ethical content.

The checklist is likely to be good at evoking reflection that expands beyond the ethical issues usually considered. For instance, questions regarding gender or international relations may initially not occur to most users, but may be spurred on by this list. Because of its abstraction, the list would be most useful if it was operationalised for HCE in advance. When operationalised for HCE it would provide an alternative structuring of issues to the approaches that are based on the Beauchamp and Childress' four principles. If such operationalisation is to be recommended, it should be clear that this approach offers some benefits over and above the other approaches. But whereas Beauchamp and Childress' principles refer to the main ethical theories, the nine checklist points are less clearly related to a systematic ethical account. In this sense, the Ethical TA appears less justified than the approaches based on Beauchamp and Childress (including the ethical matrix and the Ethical Impact Assessment approach). Also, it appears less user-friendly in its combination of abstractness and potentially too broad scope.

5.5   Ethical Constructive Technology Assessment

eCTA is a framework that provides for tools to go beyond a checklist approach, and departs from the idea that technology, society and ethics coevolve. Their methods suggests a framework for the accompaniment rather than assessment of novel technologies, since it assumes that one cannot step out of this trifold coevolution to assess moral criteria in isolation. It focuses on local contexts and on process guidance and is mostly targeted to one type of user, namely the technology developer.

With regard to this user, the approach is well-justified from the perspective of Responsible Research and Innovation and the notion of mid-stream modulation (Fisher et al. 2006). However, it is a relatively demanding process for the technology developer, involving deliberation with stakeholders over time. Moreover, the approach does not provide many ethical resources with which to guide or assist the technology developer and stakeholders to reflect ethically. Thus even if this approach could potentially incorporate a wide variety of ethical issues, it is highly uncertain that it would result in such a comprehensive ethical assessment since the involved parties are not prompted on specific ethical issues. It thus appears that the approach is valuable, but not very user-friendly and transparent, and cannot demonstrate comprehensiveness with regard to ethical concerns.

5.6   Ethical Impact Assessment

Many of the ethical issues and values grouped under the four principles on which the EIA framework is based are complementary to those encountered in the HCE ethical debate. Indeed, the EIA framework is very transparent and comprehensive in terms of listing the most important ethical issues and concerns. The framework is also useful as a means of analysing and evaluating specific applications. Moreover, the process of an EIA is clearly described in a number of steps which can be easily followed—and crucially, adapted—by the user according to application specific factors, contextual conditions and assessment purpose. As the framework is so well-developed, the user with fewer resources may use it as an analytic framework as it is. Users with greater means may use EIA as a part of a stakeholder dialogue. This flexibility is important in a field such as HCE where there are decision makers on many levels, ranging from parents to university committees or government agencies.

The complexity of the framework may appear intimidating to a user, but this may be alleviated by the fact that the questions included are of a very concrete nature. However, in order to be applicable to HCE, the framework requires a degree of adaptation and we provide such an adapted version in Table 5.2.

Basically, the decision maker may apply this version by himself or herself more or less as it is. Most likely, all dimensions will not be relevant for all decision makers and some can be skipped. Moreover, the decision maker will likely not have available information with which to answer all the relevant questions. In this case, it is important to note where information is missing, so that more information can be sought or so that the uncertainties of the application are clearly understood. In cases of great uncertainty, the Precautionary Principle may be applied.

Decision makers with more resources may choose to organise a deliberative workshop in which this generic list is adapted to the specifics of the use context and expert knowledge about the issues is elicited. Another option is to commission specific studies to gather evidence on key dimensions or questions. Decision makers with less resources will try to answer the questions within their limits. All

Table 5.2 EIA for HCE issues

Respect for autonomy (right to liberty)
Does the application curtail a person's right to liberty and security in any way? If so, what measures could be taken to avoid such curtailment?
Does the application recognise and respect the right of persons to benefit from measures designed to ensure their independence, social and occupational integration and participation in the life of the community?
Will the application constrain a person or curtail their freedom of movement or association? If so, what is the justification?
Does the person have a meaningful choice, i.e., are some alternatives so costly that they are not really viable alternatives? If not, what could be done to provide real choice?
Dignity
Will the application be developed and implemented in a way that recognises and respects the right of citizens to lead a life of dignity and independence and to participate in social and cultural life? If not, what changes can be made?
Is such a recognition explicitly articulated in statements to those involved in or affected by the application?
Does the application compromise or violate human dignity? For example, in the instance of non-invasive brain stimulation, can citizens (for instance children) decline to use it or, if not, what measures can be put in place to minimise or avoid comprising their dignity?
Does the application require citizens to use a technology that marks them in some way as cognitively or physically disabled? If so, can the technology be designed in a way so that it does not make them stand out in a crowd?
Does the application involve implants? If so, does it accord with the opinion of the European Group on Ethics (EGE)?
Informed consent
Will the application obtain the free and informed consent of those persons to be involved in or affected by it? If not, why not?
Will the person be informed of the nature, significance, implications and risks of the application? Are there special measures in place for children?
Will such consent be evidenced in writing, dated and signed, or otherwise marked, by that person so as to indicate his consent?
If the person is unable to sign or to mark a document so as to indicate his consent, can his consent be given orally in the presence of at least one witness and recorded in writing?
If relevant, does the consent outline the use for which data are to be collected, how the data are to be collected, instructions on how to obtain a copy of the data, a description of the mechanism to correct any erroneous data, and details of who will have access to the data?
If the individual is not able to give full informed consent (for instance, in the case of children) to use the application, will close relatives, a guardian with powers over the person's welfare or professional carers be consulted? Will written consent be obtained from the individual's legal representative and his doctor?
Can the user stop using the application at any time, without any adverse consequences?
Is use truly voluntary? For example, does the person need to use the application in order to get a service or benefit to which there is no alternative?

(continued)

Table 5.2 (continued)

Non-maleficence
Safety
Is there any risk that the application may cause any physical or psychological harm to consumers? If so, what measures can be adopted to avoid or mitigate the risk?
Have any independent studies already been carried out or, if not, are any planned which will address the safety of the application? If so, will they be made public?
To what extent is scientific or other objective evidence used in making decisions about the specific application?
Does the application affect consumer protection?
Can the information generated by the application be used in such a way as to cause unwarranted harm or disadvantage to a person or a group?
Does the project comply with the spirit of consumer legislation (e.g., Directive 93/13 on unfair terms in consumer contracts, Directive 97/7 on consumer protection in respect of distance contracts and the Directive on liability for defective products (85/374/EEC))?
Social solidarity, inclusion and exclusion
Has the producer of the application taken any steps to reach out to the excluded (i.e., those excluded from use of the Internet)? If not, what steps (if any) could be taken?
Does the application have any effects on the inclusion or exclusion of any groups?
Has a wide range of perspectives and expertise been involved in decision-making for the application?
How many and what kinds of opportunities have stakeholders and citizens had to bring up value concerns?
Isolation and substitution of human contact
Will the application replace or substitute for human contact? What will be the impact on those affected?
Is there a risk that the application may lead to greater social isolation of individuals? If so, what measures could be adopted to avoid that?
Is there a risk that use of the application will be seen as stigmatising, e.g., in distinguishing the user from other people?
Discrimination and social sorting
Does the application use profiling technologies?
Does the application facilitate social sorting?
Could the application be perceived as discriminating against any groups? If so, what measures could be taken to ensure this does not happen?
Will some groups have to pay more for certain services (e.g., insurance) than other groups?
Beneficence
Will the application provide a benefit to individuals? If so, how will individuals benefit from the application?
How certain is it that these benefits will occur? Is there information available regarding the trade-off between side-effects and improvements as a result of using the application for the purposes of enhancement?
Who benefits from the application and in what way?
Will the application improve personal safety, increase dignity, independence or a sense of freedom?

(continued)

Table 5.2 (continued)

Does the application serve broad community goals and/or values or only the goals of the user? What are these, and how are they served?

Are there alternative, less intrusive or less costly means of achieving such objectives?

What are the consequences of not proceeding with development or use of the application?

Does the application facilitate the self-expression of users?

Universal service

Will the application be made available to all citizens? When and how will this be done?

Will training be provided to those who do not (yet) have skills or knowledge to use it? Who should provide the training and under what conditions?

Will the application cost the same for users who live in remote or rural areas as for users who live in urban areas? How should a cost differential be paid?

Accessibility

Does the new application require a certain level of knowledge, e.g. of computers, that some people may not have?

Could the application be designed in a way that makes it accessible and easy to use for more people, e.g., senior citizens and/or citizens with disabilities?

Value sensitive design

Has the application been designed taking into account values such as human well being, dignity, justice, welfare, human rights, trust, autonomy and privacy?

Have the technologists and engineers discussed their application with ethicists and other experts from the social sciences to ensure value sensitive design?

Does the new application empower users?

Sustainability

Is the application economically or socially sustainable? If not, and if the application appears to offer benefits, what could be done to make it sustainable?

Should an application provided by means of a research project continue once the research funding comes to an end?

Does the application have obsolescence built in? If so, can it be justified?

Has the project manager or technology developer discussed their products with environmentalists with a view to determining how their products can be recycled or how their products can be designed to minimize impact on the environment?

Justice

Are all vulnerable groups that may be affected by its undertaking been identified?

Is the application equitable in its treatment of all groups in society? If not, how could it be made more equitable?

Does the application confer benefits on some groups but not on others? If so, how is it justified in doing so?

Do some groups have to pay more than other groups for the same application?

Is there a fair and just system for addressing failures or damages with appropriate compensation to affected stakeholders?

Equality and fairness (social justice)

Will the application be made widely available or will it be restricted to only the wealthy, powerful or technologically sophisticated?

Does the application apply to all people or only to those less powerful or unable to resist?

(continued)

Table 5.2 (continued)

If there are means of resisting the provision of personal information, are these means equally available or are they restricted to the most privileged?
Are there negative effects on those beyond the primary user and, if so, can they be adequately mediated?
If persons are treated differently, is there a rationale for differential applications, which is clear and justifiable?
Will any information gained be used in a way that could cause harm or disadvantage to the person to whom it pertains? For example, could an insurance company use the information to increase the premiums charged or to refuse cover?

users of the EIA need to be transparent about the evidence base for their judgements.

Some of the questions will function as prompts to the decision maker, for instance, to ensure that informed consent is gathered from users of the application. Other questions may address conditions for the use of the application, for instance that the user can stop using the application without any adverse consequences. Some questions may amount to ethical arguments against the application in general, for instance if it facilitates social sorting. Other questions will indicate the reasons for using the application (the beneficence dimension). Decision makers will need to explicitly prioritise arguments for and against the application, and refer to such priorities in their final conclusions on the application.

Compared to the four principles approach and the ethical matrix, the EIA's level of detail provides additional guidance. Compared to the ethical matrix, however, EIA may not bring out value trade-offs in an equally transparent way. But all in all, the EIA appears as a comprehensive, transparent and user-friendly approach, and a good alternative for an ethical framework for HCE.

5.7   Summary

From Table 5.3, we see that the three principle based approaches (the four principles approach, the ethical matrix and the EIA) appear to come out as most relevant for practical ethical decision-making guidance on HCE applications. The particular version of the principle based approach one prefers can vary, but all seem to have applicability in pluralistic governance situations, as described at the beginning of this book. Moreover, an additional quality of these frameworks is their systematic anchoring in ethical theory.

Table 5.3 Summary of evaluation of the different approaches

Framework	User-friendliness	Transparency	Comprehensiveness
Principle based ethics (four principles approach)	The popularity of the framework suggests high user-friendliness	It does not in itself provide a detailed structure for comprehensive transparency	All relevant concerns can be given a place in an adapted version of the approach
Ethical matrix	The popularity of the framework suggests high user-friendliness	The matrix structure provides transparency in balancing concerns	All relevant concerns can be given a place in the approach. The matrix structure demonstrates high comprehensiveness
Anticipatory technology ethics	Low—requires a high level of expert input	The checklist provides a transparent structure	Very comprehensive in terms of focus on different levels of ethical analysis and two stage ethical analysis which allows moral values and principles to be operationalised and cross-referenced with technology descriptions, in addition to an evaluation of the potential importance of ethical issues
Ethical technology assessment	Limited. Ethical TA has been designed specifically to inform innovators	The approach requires operationalisation before it supports transparency	Ethical TA is targeted to technologies in a development process
Ethical constructive technology assessment	eCTA provides an approach for the understanding of the interrelation between man and technology, which might be helpful to sociologists in elucidating the effects on the microscale of the introduction of novel technologies, but it cannot provide for ethical criteria beyond the advice that "user contexts should be taken into account". The framework suggested needs guidance by experts to be operationalised	This approach does not in itself provide a structure for transparency	eCTA is based in a classic STS approach in its view of technology, ethics and society as coevolving. It can only explain emerging forms of ethics, not argue for or against them. eCTA provides material for a qualitative approach to technologies, that takes into account a differentiated view on the nature of technologies

(continued)

Table 5.3 (continued)

Framework	User-friendliness	Transparency	Comprehensiveness
Ethical impact assessment	High—users can make use of the issues and questions clustered under the four principles to guide their assessment and to formulate other questions. EIA also allows for collaboration between non-experts and those with ethical and other forms of expertise	The detailed structure will allow for transparent consideration of all ethically relevant issues	Very comprehensive—sets out key issues and questions grouped under Beauchamp and Childress's four principles in a concrete and detailed way

References

Beekman, V., E. de Bakker, H. Baranzke, O. Baune, M. Deblonde, E.-M. Forsberg, R. de Graaff, H.-W. Ingensiep, J. Lassen, B. Mepham, A.P. Nielsen, S. Tomkins, E. Thorstensen, K. Millar, B. Skorupinski, F. Brom, M. Kaiser, and P. Sandøe. 2006. *Ethical bio-technology assessment tools for agriculture and food production: Final report ethical bio-TA tools (QLG6-CT-2002-02594)*. Agricultural Economics Research Institute (LEI). https://www.academia.edu/2838074/Ethical_bio-technology_assessment_tools_for_agriculture_and_food_production. Accessed September 1, 2015.

Cotton, M. 2014. Ethical matrix and agriculture. In *Encyclopedia of Food and Agricultural Ethics,* ed. P.B. Thompson, P.B., and D.M. Kaplan, 1–10. Berlin: Springer.

Fisher, E., Mahajan, R.L., and Mitcham, C. 2006. Midstream Modulation of Technology: Governance From Within. Bulletin of Science, Technology and Society, 26 (6): 485–96.

Forsberg, E.-M. 2007. Value pluralism and coherentist justification of ethical advice. *Journal of Agricultural and Environmental Ethics* 20 (1): 81–97.

Jensen, K.K., E.-M. Forsberg, C. Gamborg, K. Millar, and P. Sandøe. 2011. Facilitating ethical reflection among scientists using the ethical matrix as a tool. *Science and Engineering Ethics* 17 (3): 425–445.

Kaiser, M. 2005. Assessing ethics and animal welfare in animal biotechnology for farm production. *OiE, Scientific and Technical Review of the World Organisation for Animal Health* 24 (1): 75–87.

Kaiser, M., and E.-M. Forsberg. 2001. Assessing fisheries—Using an ethical matrix in a participatory process. *The Journal of Agricultural and Environmental Ethics* 14: 191–200.

Kaiser, M., K. Millar, E. Thorstensen, and S. Tomkins. 2007. Developing the ethical matrix as a decision support framework: GM fish as a case study. *Journal of Agricultural and Environmental Ethics* 20 (1): 65–80.

Kass, L.R. 2003. Ageless bodies, happy souls. *The New Atlantis* 1: 9–28.

Chapter 6
Final Reflections

6.1 Discussion

A few reflections on our methodology and findings are in order.

First, for this book, we searched for relevant literature by carrying out a systematic search using open search terms, disregarding those articles that were not relevant or were off topic and using the snowballing method in order to identify other relevant articles. We cannot claim that our search is complete, as there may be relevant articles that were not picked up in the search. However, the majority of articles included here have been written by key scholars and experts in the technical area (both in specific fields such as neuropharmacology and neuroscience) and in areas of ethics (including neuroethics and medical ethics). For this reason, we are reasonably confident that we have generated a thorough and up-to-date representation of key issues and debates in the field.

Second, we have not carried out a comprehensive survey of all possible ethical frameworks in the area of applied ethics, as this would be extremely extensive. The selection of frameworks presented here are key contributions in applied ethics and ethical based TA approaches in closely adjacent fields to HCE.

Third, it may be claimed that the distinction between tools addressing the opening up deliberation on technologies and tools addressing closing down decision making on applications is not so clear-cut. For instance, several of the questions (for instance questions 36–43) in the Socratic HTA-based approach (Hofmann 2016) are similar to the suggested EIA approach (in Table 5.1), and the HTA approach is similar to the ethical matrix in addressing various stakeholders' perspectives on specific principles and issues. However, although there is a certain overlap of questions, there are other differences. The opening up deliberation intends to question broader issues and should involve deliberating on the hard, principled questions about naturalness, human identity and how technologies change relations between people. The decision making level, on the other hand, must come to conclusions in the specific context of concrete applications, with

© The Author(s) 2017
E.-M. Forsberg et al., *Evaluating Ethical Frameworks for the Assessment of Human Cognitive Enhancement Applications*, SpringerBriefs in Ethics, DOI 10.1007/978-3-319-53823-5_6

specific consequences for defined groups of stakeholders, and a decision aiding ethical tool must assist in transparently weighing these consequences when making decisions that will have relatively immediate effects.

Fourth, our criteria have determined the particular frameworks that we argue have most appeal. Other criteria might have led us to make other recommendations. We have argued for these criteria, which appear justified also by the fact that other key scholars highlight the same. However, readers may disagree with the criteria. We believe that we have facilitated such critical evaluation by the reader by being transparent.

Fifth, in this book we claim, in accordance with Moula and Sandin (2015), the importance of user-friendliness. This criterion carries with it some ideas of "the user" or "the users". Kaiser et al. (2007) engaged both an expert and a lay group for their workshops testing the ethical matrix, while Millar et al. (2007) used a stakeholder approach. Likewise, Jensen et al. (2011) applied the ethical matrix in workshops comprising only of researchers and in a workshop consisting of stakeholders. These tests showed that users in different settings experienced the same tool slightly differently. From the concrete world of tools, we know that different types of tools may be used in order to complete an identical task. Some tools require extensive training and skill while other tools are more integrated in a cultural setting. Thus, in a selection of possible tools, it is challenging to stand on the outside of a social setting and state the superiority of one tool over another based on criteria concerning "the user". Flexibility of the approach related to different user needs and preferences therefore seems important. The EIA appears to have such versatility, as does the ethical matrix.

Sixth, we have argued that the cases we have discussed here represent real world applications of HCE, and not only speculative scenarios. However, one might object that even for applications already in use, their potential to supply the desired enhancement effect may still be considered speculative, since this has not yet been documented. So by proposing ethical decision guiding frameworks on HCE we contribute to cement a hype that is not anchored in reality, taking attention away from more urgent issues (for such an argument, see Nordmann and Rip 2009). We acknowledge this risk, but believe that there is a need for a rich ethical discussion about these applications even if their enhancement potential is uncertain. In fact, by engaging in such a concrete and detailed assessment, potential hyped claims may be brought to the surface. It is harder to avoid assuming enhancement potential when discussing enhancement at a more philosophical level.

Similarly, as the off-label use of prescription drugs is not legal it may be held as irrelevant to give an ethical assessment of such use. However, as such use does occur—and the use of some HCE applications indeed is unregulated—we believe that it is relevant to confront the richness of ethical aspects related to concrete HCE applications in practical decision-making by private and public decision-makers, and thus that systematic ethical tools are relevant. It may also be that alternative use of pharmaceuticals is indeed ethically acceptable (or even desirable) and the ethical assessment of the enhancement use may well point to the need for regulatory change. Again, this would be a reason for thorough and realistic ethical assessment.

Finally, checklist approaches have been criticised, for instance by Kiran et al. (2015), while we here recommend such an approach (both in the form of a matrix and as part of the EIA). The reason is that we believe a structured approach is useful for non-expert users. It should also be clear that the checklist may be modified by the user to fit with the context, so a checklist is not a straitjacket, but an instrument to prompt consideration of a range of issues. Moreover, it should be clear from the above discussion that we believe that in addition to this level of concrete decision making tools there is also a need for more open ethical debate about broader societal aspects and more long term, or futuristic, if you will, technological development in HCE. But it should not come at the cost of assessing applications we already find in our societies in a systematic and concrete way, and flexible, principle based approaches are appropriate for this purpose.

6.2 Conclusion

We believe that it is crucial to have a societal discussion on ethical issues of HCE technologies and policies in general, as well as a scholarly systematic deliberation on overall ethical and philosophical issues, as outlined by Hofmann (2016). However, we have argued for the need also for decision guiding approaches more targeted to specific applications in specific contexts; assisting both public and private decision makers in making regulatory, funding or commercial decisions on specific applications.

From our discussion of the ability of six different frameworks to handle the most urgent HCE related ethical issues in a transparent and user-friendly way, we have proposed the ethical matrix and ethical impact assessment (both building on the four principles framework) as the most appropriate ethical decision guiding approaches within the scope of this study. We have developed a version of the ethical matrix and the EIA, adapted to HCE.

This does not mean that better frameworks cannot be found, now or in the future, but it means that of the ones that are commonly used in the biomedical, biotechnology and ICT fields these frameworks appears to fit well to practical ethical decision guidance for HCE applications.

We have shown that these frameworks can incorporate the ethical issues and concerns proposed in the literature on the two selected HCE cases. However, we believe that they could fit other HCE cases as well, as many of the ethical issues are generic for the field.

Ideally, a database for learning should be generated as ethically informed decisions are made in this field. This should include information about how the principles and specifications are applied and balanced, and the kinds of challenges users experience in gathering sufficient evidence. This would also allow for refinement of the frameworks suggested here.

As HCE applications will continue to enter the market, ethical and governance resources should correspondingly be developed. This book has been an attempt to contribute to this aim.

References

Hofmann, B. 2016. Toward a method for exposing and elucidating ethical issues with human cognitive enhancement technologies. *Science and Engineering Ethics.* doi:10.1007/s11948-016-9791-0.

Jensen, K.K., E.-M. Forsberg, C. Gamborg, K. Millar, and P. Sandøe. 2011. Facilitating ethical reflection among scientists using the ethical matrix as a tool. *Science and Engineering Ethics* 17 (3): 425–445.

Kaiser, M., K. Millar, E. Thorstensen, and S. Tomkins 2007. Developing the ethical matrix as a decision support framework: GM fish as a case study. *Journal of Agricultural and Environmental Ethics* 20(1): 65–80.

Kiran, A.H., N. Oudshoorn, and P.-P. Verbeek. 2015. Beyond checklists: Toward an ethical-constructive technology assessment. *Journal of Responsible Innovation* 2 (1): 5–19.

Millar, K., E. Thorstensen, S. Tomkins, B. Mepham, and M. Kaiser. 2007. Developing the ethical delphi. *Journal of Agricultural and Environmental Ethics* 20 (1): 53–63.

Moula, P., and P. Sandin. 2015. Evaluating ethical tools. *Metaphilosophy* 46 (2): 263–279.

Nordmann, A., and A. Rip. 2009. Mind the gap revisited. *Nature Nanotechnology* 4: 4–273.

Uncited References

COMECE. 2008. Ethical questions raised by nanomedicine. *Science and Ethics. Opinions Elaborated by the Bioethics Reflexion Group* 1: 23–28.

Dunlop, M., and J. Savulescu. 2014. Distributive justice and cognitive enhancement in lower, normal intelligence. *Monash Bioethics Review* 32 (3): 189–204.

Eckenwiler, Lisa A., and Felicia G. Cohn (eds.). 2009. *The ethics of bioethics: Mapping the moral landscape.* Baltimore: JHU Press.

Forlini, C., J. Schildmann, P. Roser, R. Beranek and J. Vollmann. 2015. Knowledge, experiences and views of German University students toward neuroenhancement: An empirical-ethical analysis. *Neuroethics* 8 (2): 83–92.

Grin, J. and A. Grunwald (eds.). 2000. Vision assessment: shaping technology in 21st century society—Towards a repertoire for technology assessment. *Wissenschaftsethik und Technikfolgenbeurteilung* 4.

Ida, R. 2009. Should we improve human nature? An interrogation from an Asian perspective. In *Human Enhancment*, ed. J. Savulescu and N. Bostrom. Oxford: Oxford University Press.

Ott, R., and N. Biller-Andorno. 2014. Neuroenhancement among Swiss students—A comparison of users and non-users. *Pharamcopsychiatry* 47: 22–28.